Weed Biology and Climate Change

Weed Biology and Climate Change

LEWIS H. ZISKA, Ph.D.
Plant Physiologist
Crop Systems and Global Change Laboratory, USDA
Agricultural Research Service
Building 1, 10300 Baltimore Avenue
Beltsville, MD 20705

JEFFREY S. DUKES, Ph.D.
Associate Professor
Department of Forestry and Natural Resources, and
Department of Biological Sciences
Purdue University, 715 West State Street
West Lafayette, IN 47907

WILEY-BLACKWELL
A John Wiley & Sons, Ltd., Publication

Edition first published 2011

Blackwell Publishing was acquired by John Wiley & Sons in February 2007. Blackwell's publishing program has been merged with Wiley's global Scientific, Technical, and Medical business to form Wiley-Blackwell.

Editorial Office
2121 State Avenue, Ames, Iowa 50014-8300, USA

For details of our global editorial offices, for customer services, and for information about how to apply for permission to reuse the copyright material in this book, please see our Website at www.wiley.com/wiley-blackwell.

Library of Congress Cataloging-in-Publication Data

Ziska, Lewis H.
Weed biology and climate change / Lewis H. Ziska, Jeffrey S. Dukes.
p. cm.
Includes bibliographical references and index.
ISBN 978-0-8138-1417-9 (hardcover : alk. paper)
1. Weeds–Climatic factors. 2. Weeds–Control. I. Dukes, Jeffrey S. II. Title.
SB611.Z57 2011
581.6′52–dc22

2010027035

A catalog record for this book is available from the U.S. Library of Congress.

Set in 10.5/13 pt Times by Aptara® Inc., New Delhi, India

1 2011

This book is dedicated to

Dr. David T. Patterson and Dr. James A. Bunce.
Scholars *and* Gentlemen.

Preface

Ask people for an example of a weed, and you are likely to get a variety of responses. For most suburban commuters, dandelion, chickweed, or crabgrass and their effect on the aesthetic quality of their lawn might be a concern. For avid gardeners, mustard, plantain, or clover might be a consideration. Farmers may worry most about a different group of weeds, many of which are wild relatives of the crops they grow. For a hiker or backpacker, or an ecologist, invasive plant species such as cheatgrass or yellow starthistle would top the list.

But there are other weeds that people rarely think about. For example, the allergist works with pollen outbreaks from unregulated ragweed proliferation. A dermatologist may deal with contact dermatitis from poison ivy exposure. A pharmacist might study sweet Annie, a common roadside weed in Virginia for the production of artemesinin, a new antimalarial drug. Drug Enforcement Agency (DEA) agents could study wild poppy production and its impact on the global supply of Heroin.

People rarely stop to think about the full impact of weeds in human society. Weeds affect many aspects of our daily lives, from the quantity and quality of the foods we enjoy, to the medicine we consume, the allergies we encounter, the frequency of fires in some parts of the world, and even the diversity of species in natural ecosystems.

There are, of course, existing tomes that assess and quantify all of these aspects of weed biology; the goal of this book is not to duplicate previous efforts. Rather, we attempt to assess and synthesize recent information regarding the unprecedented increase in atmospheric carbon dioxide and the associated change in climate, particularly temperature and precipitation, in regard to basic and applied aspects of weed biology. We examine responses of weed growth and reproduction as well as weed management, invasive weeds, public health, ecosystem functioning, and food security.

There is still much to be learned about the responses of weeds to anthropogenic increases in CO_2 and climate change. However, we are now beginning to recognize the scale and rapid pace of environmental perturbations caused by human society, and the implications of these changes for weed biology. Given this pace, and the interactive dynamic between weeds and environmental disturbance, we feel there is a critical need to begin summarizing and publicizing what is currently known in this field.

We begin with separate overviews of weed biology and climate change, and then discuss probable responses to climatic change and rising carbon dioxide, from the genome to the plant community level. We hope to provide a framework for greater insight into the likely impacts of global change on agronomic weeds, invasive plant species in natural or less managed plant communities, and the links between weeds and public health. We discuss the implications of global change with respect to chemical, cultural, biological, and mechanical weed management as well as some beneficial aspects of weeds. We attempt to highlight crucial research questions and to highlight adaptation or mitigation strategies that could reduce potential consequences.

We stress that this is only a beginning. We are seeking new ideas and models that can help us define, understand, and predict weed biology and likely societal impacts of unparalleled human-induced climatic uncertainty in the twenty-first century. In doing so, we wish to cast our academic net not only within the weed science community, but also among a wide swath of stakeholders, including environmentalists, agronomists, and health care providers.

To that end, we are indebted to Justin Jeffryes and Shelby Allen of Wiley-Blackwell publishing for their invitation to prepare this book and for their advice and encouragement. We are also grateful to our own scientific mentors, Hal Mooney at Stanford University, Jim Ehleringer at the University of Utah, Chris Field at the Carnegie Institution for Science, Paul Epstein at Harvard, David Patterson at Duke, Jim Bunce at USDA, and Tony Hall at University of California, Riverside, who have fostered our appreciation of nature's wonders and provided a healthy dash of skepticism. We are also indebted to our colleagues who gave generously of their time in helping put this book together, Jodie Holt, Jil Swearingen, Bethany Bradley, Dana Blumenthal, Jack Morgan, and Hilda Diaz-Soltero. Finally, we express our profound gratitude to our families and friends, who, while supporting us, must have wondered how grown men could be so fascinated with weeds.

Lewis H. Ziska, Ph.D. USDA-ARS
Crop Systems and Global Change Lab

Jeffrey S. Dukes, Ph.D. Associate Professor,
Purdue University

1 A Brief History of Weeds and Their Impact

"... the sun never sets on the empire of the dandelion"

Alfred W. Crosby, *Ecological Imperialism*

Weed Classification

Weeds were, and are, the largest single limitation on crop yield. Before the onset of chemical control for weeds, most of the work on the farm from June through August was hoeing; a weed control method still practiced around the world. Without mechanization, the size of a farmer's holding and yield was determined by how well (and how fast) a family could weed its land. As suggested by Zimdahl (1993), more human labor may be expended on weeding than on any other human enterprise (Figure 1.1).

Weeds affect nearly every aspect of our lives, from the appearance of our lawns to the quality of the food we eat, even the state of our health. Weed science, a subset of botany, is a multidisciplinary science with the goal of a systematic understanding of both weed biology, and more importantly, weed control. Weed scientists represent a broad spectrum of scientific disciplines including plant physiologists, botanists, agronomists, chemists, molecular biologists, biochemists, and ecologists.

But what makes a plant a weed? If we called something a weed in the northeastern United States, would it still be considered a weed in Australia? Are there common characteristics among different weed species? In what habitats are weeds found? How do weeds spread? What makes a weed harmful? How do we measure harm? How do we manage weeds currently? How will we manage them in the future? In this first chapter we will explore these basic concepts of weed biology.

A weed is a weed is a weed? One of the greatest difficulties in introducing the subject of weed biology is a clear understanding of what is meant by "weed." The term "weed" does not exist in nature; rather, plant species assume the mantle of "weed" when classified as such by human society.

Given the diversity of societies, many weed science books will list a number of cultural and scientifically derived definitions. While varied in scope, these characterizations fall into two broad categories: "an unwanted or undesired plant species" and "early vegetation following a soil disturbance." The first classification is used by many laypersons as well as agronomists and weed scientists, the latter is a definition widely accepted by ecologists and

Figure 1.1 The author's (LHZ) future in-laws working on a farm near Samantha, Alabama, during the early 1900s. Farm work, particularly hoeing weeds, was the main activity during the summer months, and the principal reason school children were given summers off. (Photo courtesy of Mollie Guy of Tuscaloosa, Alabama.)

environmentalists. There is also a third definition—that a weed is a plant whose virtues have yet to be discovered—suggesting that if a use were found for a weed species, it would cease to be a weed. Yet, many of the most pernicious weeds were introduced specifically because of their virtues (e.g., forage crops such as Johnson grass[1] and kudzu, as well as aesthetic species such as purple loosestrife).

Because the characterization of a weed is determined by human activity, the definition of a weed is fluid; it can vary by time and space, and certainly by culture. For example, jimsonweed, in addition to being a common weed of roadsides, is also considered an important medicinal plant for native peoples of the southwestern United States. However, certain plants are universally reviled because it is acknowledged among disparate regions and cultures that those plants can cause widespread economic or environmental damage. One such plant is Canada thistle, considered among the worst weeds in North America (Skinner et al., 2000) for its ability to invade, colonize, and out-compete native and agricultural vegetation. In

[1] Common names of weeds will be used throughout the book, scientific names, Bayer and WSSA codes are referred to in the appendix

these cases, the definition of a weed is less in doubt because the impact of these plants is universally recognized.

Ultimately, the definition of "weed" is anthrocentric and therefore flexible; however, "weeds" in many societies share certain biological characteristics.

The following is a synthesis of some of these characteristics, most notably those presented in Zimdahl (1993) and Radosevich et al. (1997). No single weed possesses all of these attributes, but these characteristics encompass biological traits associated with "weediness."

1. Long-term seed survival in the soil, including resistance to microbial degradation. In addition, weed seeds can germinate over a wide range of environmental conditions. Soil disturbance may be needed for germination.
2. Rapid growth following germination.
3. Mechanical (e.g., spines) and chemical (e.g., poison) barriers to herbivory during vegetative growth.
4. Grow over a wide range of environmental conditions, including extremes of temperature, water, and nutrient availability; high environmental plasticity.
5. Flower early during vegetative growth. Time from flowering to seed production can be rapid (e.g., 2 weeks from flowering to mature seed for Canada thistle).
6. While some weeds are self-compatible, most cross-fertilize between individuals, usually using unspecialized pollen carriers or wind. As a consequence, seeds produced by any one plant can be very genetically diverse.
7. Weeds may produce more than one seed crop per year. Individuals can produce tens of thousands of seed in a given year.
8. Reproduction, particularly for perennial weeds, may include both sexual (floral) reproduction, but also asexual reproduction from roots and/or rhizomes.
9. Weeds may have highly effective seed dispersal mechanisms, from wind (dandelion) to animals (cocklebur).

Life Cycle

Weeds can also be classified on the basis of their life history. For instance, knowing whether a plant is an annual, biennial, or perennial can help weed scientists determine where a weed could become a problem.

An annual is a plant that completes its entire life cycle (from seed germination to seed development) in less than 1 year. Annuals grow quickly, and many are prolific seed producers. Summer annuals germinate in spring, grow in summer, flower, and die with the onset of frost in the fall. Examples of summer annual weeds include velvetleaf, pigweeds, and foxtail grasses. Winter annuals, in contrast, germinate in fall or early winter, flowering and maturing in spring or early summer of the following year. Examples of winter annual weeds include chickweed, shepherdspurse, and cheatgrass.

Biennials, as the name suggests, are plant species that complete their life cycle over a 2-year period. These species typically form a vegetative rosette during the first year of growth, become quiescent over winter, and then bolt during the following spring, forming a tall

Figure 1.2 Chickweed, curly dock, and field bindweed as common examples of annual, biennial, and perennial weeds.

inflorescence with prodigious seed development. Examples of biennials include curly dock, garlic mustard, bull thistle, and musk thistle.

Among the most damaging of weeds are perennials, those that are long-lived, multi-year plant species. These can be divided into two subgroups: simple and creeping (Zimdahl, 1993). Simple perennials reproduce and spread primarily by seed, whereas creeping perennials can spread by both sexual (i.e., seeds) and asexual (i.e., rhizomes and roots) means. Examples of simple perennials would include common lawn weeds such as dandelion or plantain, or weeds of wet damp locations such as curly dock. Creeping perennials include weeds associated with contact dermatitis such as poison ivy (rhizomes and stems), and common weeds in pastures such as leafy spurge (rhizomes) and Canada thistle (roots) (Figure 1.2).

Classification by Physiology

Differences in photosynthetic pathway provide another means to classify weeds. This type of classification will also be important when we examine ongoing increases in atmospheric carbon dioxide associated with climate change in later chapters.

Initial investigations of how plants acquire carbon led to the discovery that one of the first products of photosynthesis was the production of phosphoglyceric acid, a three-carbon compound (Calvin, 1949). For the vast majority (approximately 95%) of plant species it was discovered that this type of photosynthesis (termed C_3 because the first stable product had three carbons) was the sole means by which carbohydrates were synthesized from light, energy and water. However, in the 1960s, researchers discovered other photosynthetic pathways, most notably the C_4 pathway, where the first products of photosynthesis are four-carbon sugars such as oxaloacetate, malate, and aspartate. Although these C_4 plants, principally tropical grasses, constitute only about 4% of all plant species, a higher percentage of this type of photosynthesis appears to occur among the world's worst weeds (e.g., Holm et al., 1977). A third type of photosynthetic pathway, crassulean acid metabolism or CAM, occurs in less than 1% of plant species. While succulent species use CAM photosynthesis, major weeds do not.

Classification by Habitat

One of the most well-recognized habitats for weeds is cropland. However, because crops cover a wide range of different environments, there may not be a specific set of characteristics associated with crop weeds per se. Interestingly, the most competitive weeds in a given cropping situation are often simply "imitators" or wild relatives of the crop. They are selected for by cultural practices such as planting date, fertilizer application, selective herbicide usage, etc. which while favoring the crop, also select for those weeds that are closely related physiologically, morphologically, or phenologically. Examples of crops and their weed relatives include potatoes and nightshade (both in the genus *Solanum*), rice and wild or red rice, sorghum and shattercane (both *Sorghum bicolor*), oat and wild oat.

Rangelands constitute those land areas characterized by grassland and shrubland used (although not exclusively) for animal grazing. Because of their use, classifications in rangelands are associated with those weeds whose presence is either directly (e.g., poisonous) or indirectly (competes with desired grasses) harmful to large animal (sheep and cattle) grazing. Examples of typical rangeland species include leafy spurge and yellow starthistle.

Forests have another unique set of weed species associated with them. James Miller of the U.S. Forest Service has identified over 33 plants or groups that are spreading rapidly through southern forests (Miller, 2003). As might be expected, vines that compete directly for sunlight such as kudzu, Oriental bittersweet, or English ivy are considered among the worst nonnative vines, but other perennial weeds such as Chinese and Japanese privet, nonnative roses, and sacred bamboo are also among the worst forest weeds. In addition, certain tree species can also be considered weeds and among these, nonnative trees such as Princess tree and tree of heaven are considered common weeds in forest environments.

Riparian weeds are those species that are found in wet, poor drainage areas. They are adapted to anaerobic soils and may rely on water as a means to spread by seed or asexually. Riparian weeds include purple loosestrife, curly dock, giant reed, or salt cedar.

Aquatic weeds are those weeds adapted to living directly in the water. They can be floating, submersed, or immersed (submersed but with a root system). They can reduce water quality and water supply as well as pose problems for navigation or power generation. Some of the most prevalent aquatic weeds include water hyacinth and hydrilla (Figure 1.3).

A Few Basic Principles of Weed Ecology

Weed ecology is that aspect of weed biology that studies weed distribution, growth, development, reproduction, and population dynamics of weed species in managed (human) and unmanaged (natural) plant communities.

Figure 1.3 Weeds by habitat: (1) common lambsquarters in a soybean field; (2) Canada thistle as a rangeland weed in South Dakota; (3) tree of Heaven, a common weed in urban areas; (4) purple loosestrife, a common riparian weed in marshes; (5) water hyacinth, a common aquatic weed. (Photos are all from USDA-ARS Image Gallery.)

Seed Biology

We can begin with seed, and an appreciation of the ability of weeds to produce them. The number of weed seeds present in soil can be enormous. Koch (1969) has derived figures that indicate between 30,000 and 350,000 weed species per square meter (or 300 million to 3.5 billion seeds per hectare). An estimate of the number of seeds produced for a single weed can be in the hundreds of thousands (Table 1.1).

What is the fate of seed once it enters the soil? Harper (1977) envisioned the soil as a seed "bank" in which "deposits" or "withdrawals" are made. An example of a deposit would be seed rain, whereas withdrawals (seed removal) would occur by germination, deterioration, or predation. The concept of a seed bank is a very useful one for weed ecology since it provides key information regarding seed diversity and longevity as well as long-term information as to efficacy of weed management.

Seed longevity in turn, will depend on a number of factors and their interaction. Obviously, how much seed a given species produces is an important consideration as is soil type. However, seed decomposition, predation, and dormancy/germination are also key considerations. Overall, observations indicate that while seeds from grass and crop species are rarely long-lasting within the seed bank, weed seeds may last for hundreds of years if not millennia (Odum, 1965).Weed seeds are almost always present in the soil.

Seed germination and emergence is dependent on dormancy. Although there is a recognized physiological aspect of dormancy (seed ripening), most dormancy is imposed by the environment. Dormancy, in turn, can be broken by a given environmental factor or combination of factors, usually those that occur in seasonal cycles. For example, freeze/thaw combinations may be necessary to remove seed coat restrictions. Alternatively, many weed seeds need light (Sauer and Struik, 1964), which is usually abundant following a physical disturbance in the environment.

Table 1.1 Seed production (per plant) and qualification (+ or −) as to the presence of asexual reproduction for selected weed species.

	Seed	Asexual	Comments
Canada thistle (*Cirsium arvense* [L.] Scop.)	530–5,300	++	Extensive below ground root system 3 m horizontal and vertical lengths
Common lambsquarters (*Chenopodium album* L.)	9,000–370,000	−	Prolific seed producer
Common mullein (*Verbascum thapsus* L.)	200,000–300,000	−	Prolific seed producer
Common ragweed (*Ambrosia artemisiifolia* L.)	3,000–5,000	−	Agronomic weed and public health issue
Dalmation toadflax (*Linaria dalmatica* L. [Mill])	up to 500,000	++	Extensive deep root system
Diffuse knapweed (*Centaurea diffusa*)	up to 18,000	−	Not asexual, but can regrow from cut crown
Field bindweed (*Convolvulus arvensis* L.)	25–300	++	Extensive rhizome growth and establishment of large colonies
Musk thistle (*Carduus nutans* L.)	up to 100,000	−	Normally biennial, can be annual with warmer temperatures
Perennial sowthistle (*Sonchus arvensis* L.)	up to 13,000	++	Can spread extensively by roots
Purple loosestrife (*Lythrum salicaria* L.)	up to 3,000,000	−	Can regrow from woody rootstock if shoots are cut
Quackgrass (*Elymus repens* L.)	up to 8,000	++	Extensive rhizome production, can form large colonies
Wild oat (*Avena fatua* L.)	200–700	−	Common weed in cereal production
Yellow starthistle (*Centaurea solstitialis* L.)	up to 150,000	−	Can regrow if shoots cut

Reference : www.invasive.org.

Seed germination is associated with rapid metabolic activity, radicle emergence, and then shoot appearance. Germination is a perilous ecological proposition, and natural selection has caused the timing of germination to coincide with favorable environmental conditions (Probert, 1992). However, even with such selection, given the large seed populations of some weed species, it is evident that many of them fail to germinate, decompose, or get eaten.

What then are "favorable environmental conditions" that promote weed seed germination within the seed bank? Whether human or natural, any disturbance exposes soil to light, alters water and temperature regimes, and, assuming temperatures above thermal limits, can allow seeds to break dormancy and germinate. As we have seen, weeds can be a primary constituent of the seed bank; such physical (abiotic) disturbances are therefore a key factor in their establishment within the plant community.

Emergence and Competition

Given physical disturbance of the soil as a signal for extensive seed germination, what factors following emergence from the soil are associated with weed species success? What, in other words, makes some weeds more competitive than others?

It is difficult to provide a complete answer to these questions, and scientists are still addressing many aspects of competition. However, in general, success at competition reflects the ability of a given species to acquire a limited set of resources at a greater rate than its neighbors, whether those neighbors are a different species (interspecific competition) or the same species (intraspecific competition). How these resources are acquired depends on both genetic hard-wiring (e.g., rate of leaf development, nitrogen use efficiency, etc.) and spatial and temporal variability of physical resources (e.g., nitrogen distribution in the soil, rainfall, temperature, etc.). In general, plant competition is for abiotic resources, principally light, water, and nutrients between or within species.

Because light varies in duration, intensity, and quality, and must be used immediately or lost permanently, those plants that intercept light and shade their neighbors are at a competitive advantage. Plants invest heavily into new leaf formation, particularly after emergence. Indeed, a simple comparison of leaf area ratio among seedlings can be used as one measure of early competitive success between weeds and crops (Kropff and Spitters, 1991). However, too much investment in leaves at the expense of stems or branches may also result in a disadvantage, since height is an important consideration in light interception. Overall, in many cropping situations where water and nutrients may be nonlimiting, light competition may be one of the principal factors influencing weed–crop interactions (Donald, 1961) and reducing crop productivity (e.g., oat and wild oat; Cudney et al., 1991).

Water availability varies greatly by region, and competition for water among seedlings can be severe in areas where natural rainfall is low or inconsistent (e.g., rangelands and savannah). Less competition is usually associated with irrigated systems, although variation in delivery times and water quality (e.g., salinity) may be an additional feature related to competitive success. Overall, there are three factors associated with competition for water as suggested by Radosevich et al. (1997): (a) variation in the total amount and temporal availability of water inputs into the soil; (b) soil volume available to the plant, i.e., the rate of root development and exploration of that volume; and (c) physiological factors related to the exchange between carbon acquisition and water loss. For some weedy species, rapid depletion of available water may be an effective strategy for competition with crops (e.g., cocklebur and soybean; Scott and Geddes, 1979); alternatively, a number of common weedy species have the C_4 photosynthetic pathway, which increases overall water use efficiency (the ratio of CO_2 uptake to H_2O loss) and may be at an advantage within the plant community if water becomes limiting. In general, success in competition for water will be dependent on the relative abilities of the individual species to obtain the resource and to tolerate water deficits (Table 1.1; Patterson, 1995).

Nutrients are a recognized environmental resource needed for plants to complete their life cycle. Nutrients are divided into two broad categories based on the relative needs of the plant, namely, macronutrients (those needed in large amounts), such as carbon, oxygen, hydrogen, nitrogen, potassium, calcium, phosphorous, magnesium, and sulfur; and micronutrients (needed in lesser amounts) such as iron, chlorine, copper, manganese, zinc, molybdenum, and boron. Competition for these nutrients is both dynamic and complex since nutrient availability is dependent on soil characteristics, most notably pH, and aspects of plant morphology such as root development, that are species specific (Patterson, 1995). It is difficult then to predict a precise competitive outcome for an explicit nutrient deficiency. Nevertheless, there are some generalizations that can be made. First, it is clear that weeds utilize nutrient resources at least as rapidly as many crop species. This suggests that weed removal should result in greater nutrient availability to the crop species. Second, there is evidence that weeds may be more efficient at removing nutrients from the soil than cultivated plant species (Vengris et al., 1955).

The previous examples are meant to be illustrative, and are by no means a complete synthesis of what is known regarding competition. For example, while nutrients, water, and light are presented as individual physical limitations, it is common for one or more to be limiting simultaneously. In addition, most competition experiments have focused on suboptimal conditions, not on circumstances where such factors may be in excess (e.g., flooding and agronomic fertilization). Yet, differential responses to an overabundance of resources may also have significant implications for plant–plant interactions.

Weed Reproduction and Spread

Having examined how weeds survive in the seed bank, germinate, emerge, and compete successfully, we can turn our attention to reproduction and distribution within the environment. How do weeds become distributed in the environment?

One reproductive method, common among many perennial weeds, is through vegetative or asexual reproduction. Vegetative portions of the plant capable of regeneration include stolons, long, slender stems that grow parallel to the soil surface (e.g., Bermuda grass); rhizomes, stems that grow underground, while producing adventitious roots and shoots (e.g., quackgrass and poison ivy); tubers, which are extensive belowground storage organs (enlarged rhizomes) that possess axillary buds (e.g., Jerusalem artichoke); bulbs, modified underground buds capable of regeneration (e.g., wild garlic and wild onion); corms, enlarged underground perennial stems for carbohydrate storage (e.g., wild onion); and roots (usually horizontal that can give rise to separate shoots (e.g., Canada thistle and field bindweed). Although many perennials do produce vegetatively, most, if not all, also reproduce sexually through flowering. Hence, such weeds can maintain both a high density of individuals for a given surface area, as well as spread via seed dispersal. It is important to distinguish the seed bank and the presence of vegetative bud reserves since both are an aspect related to establishment, dominance, and competition from the soil. In general, buds are representative clones from a single plant, whereas the population of buried seeds represents a broader range of genotypes. Consequently, vegetative reproduction should be at an advantage when environmental conditions are stable, whereas seed genotypes would be favored with continuous disturbance. This would suggest that after a soil disturbance, seed genotypes would dominate initially, followed by the appearance of herbaceous perennial weeds with high vegetative bud reserves. Such vegetative reserves can be impressive. For example, within 2 years, Canada thistle plants can produce over 66 feet (20 m) of new roots (Parsons, 1992).

Getting the Seed Out

As we have already observed, one of the characteristics of weed species is prodigious seed production. However, widespread distribution of these seeds, through a wide array of vectors, aids in their spread and negatively impacts their control.

A number of weed species have plumed seed for wind dispersal. These include dandelion, Canada thistle, mare's tail, and a number of weeds within Asteraceae. The plume, or pappus, is the "parachute" used for wind dispersal, with the extent of wind displacement related to the ratio of seed weight to pappus size. That is, those species that produce a low ratio of achene diameter to pappus size have higher "hang-time" and can disperse farther from the mother plant. Plant size, particularly the height of seed release, is also positively correlated with area of dispersal. Tall plants (those over 1 m in height) can distribute seed as far as 5 km (Dauer et al., 2007) depending on wind velocity and degree of infestation. In addition to disseminating seed directly, wind can also move senescent or detached portions of a plant, with subsequent distribution of seed over a wide area. For example, tumbleweeds such as Russian thistle can travel several kilometers while dispersing 60,000 seeds per plant (Stallings et al., 1995).

A number of weed seeds without any special morphological modifications can travel by water. This is a particular concern in agricultural areas where irrigation is common. An early study by Eddington and Robbins (1920) found a total of 81 different weed species in irrigation ditches in Colorado. A study done by Wilson (1980) found a total of 77 different plant species in three main irrigation canals over two seasons, collecting a total of 30,346 seeds. Interestingly, many weed seeds are able to survive long periods of time submerged in water. For example, seed of field bind weed had 55% germination, even after 54 months under water (see Table 5.3; Zimdahl, 1993). Water then, can be an effective means to distribute weed species over a large geographic area.

Many weed seeds are characterized by having various hooks and/or barbs on the seed coat (e.g., puncturevine and cocklebur). Others may possess a sticky coating or resin (e.g., gumweed). These features are commonly associated with mechanical dispersal of weed seeds by animals. Spread of some weed species in the western United States, for example, can be traced to grazing areas of sheep and cattle (Crafts, 1975). In addition to such mechanical dispersal, animals may also distribute seed directly by ingestion followed by partial digestion. After passing through the alimentary canal, seed can then be spread via feces. Harper (1977) and King (1966) provide a number of examples of seed dispersal by birds, rodents, and wild and domesticated ruminants. Bird distribution in particular can move seeds over thousands of miles (Proctor, 1968).

Human activities, particularly in agriculture, are a well-recognized vector in weed seed dispersal. Farm machinery, such as plows, harrows, and even tires can spread weed seed well beyond its original source. Farming practices influence which weed species become established. For example, in rice farming, direct seeded rice is associated with the spread of wild or "red" rice because the seed of red rice is often indistinguishable from that of the cultivar. Conversely, rice that is transplanted into flooded conditions is associated with weeds that are tolerant of such wet or paddy environments such as *Echinochloa* species.

In addition to farming practices, the globalization of trade has also resulted in the introduction of numerous "exotic" or "invasive" weed species. For example, Japanese stiltgrass, a forest understory weed, was introduced as packing material into Tennessee in the early twentieth century (Fairbrothers and Gray, 1972); cheatgrass may have been introduced as a contaminant with crop seed (Novak and Mack, 2001). Introduction of other weedy species has been deliberate because their initial introduction was thought to be of economic benefit. Perhaps the most notorious example is kudzu, a leguminous perennial vine; now spread over 8 million acres in the southeastern United States, whose properties as forage and feed for cattle attracted the attention of the USDA, which heavily promoted the planting of kudzu as late as the 1940s.

The Harm that Weeds Do

So far we have discussed a number of principles that can be used to characterize weeds and their ecology. This may seem like a pointless exercise, since many people do not consider plants to be "bad." However, it is important to acknowledge the harmful nature of weeds and to recognize that "green" is not always good.

Food

Agriculture is the principal provider of food for the human population. Of the 250,000+ plant species on Earth, only a few species have been able to be specifically cultivated for human consumption, and of those, three cereals, rice, wheat, and corn supply the global population with approximately 50% of their calories. Overall, more than 90% of the population is supplied by less than 20 plant species (Diamond, 1997). Although one can argue the benefits of modern agriculture (ability to feed large numbers with small amounts of land) versus its detriments (large inputs of fertilizer and water contaminants), there is no question that the current population of the world (6.9 billion) is very much dependent on modern agricultural techniques and the maintenance of high crop yields.

Any impediment on these yields therefore has enormous implications for the ability of a region or country to provide sufficient food for its population. The three major biological barriers to maximal crop yields are insect pests, weeds, and plant pathogens. Among these, weeds are the largest limitation to maintaining crop yields. There are many times more weed species than crops; for example, in the United States, there are approximately 10 weed species that compete with each crop species (Bridges, 1992).

Weed/crop competition is a major factor limiting food production worldwide. For example, in China 10 million metric tons of rice is estimated to be lost annually due to weed competition (Zhang, 2001), enough rice to feed 56 million people for 1 year (Labrada, 2007).

In addition to their direct effects on reducing crop yields, some weeds host damaging insects or pathogens. Tall morning glory, for example, is a host for leaf blight, cocklebur for stem rust; insects such as the corn borer may use Russian thistle as a means to spread curly top. Other weeds may be hosts for different nematodes that damage soybean and corn.

While negatively impacting yields, weeds may also affect quality. The presence of wild oat or red (weedy) rice can reduce milling quality of rice and wheat flour, and can reduce brewing quality for beer. The presence of weeds within forage crops such as hay can also reduce both the protein concentration and the economic return for growers.

Negative effects of weeds on pasture and forage crops can also reduce animal quality. In western rangelands, some widespread weed species such as larkspur, leafy spurge, and yellow starthistle are poisonous or do direct physical damage to animal grazers. Such damage is considerable. For example, it has been estimated that leafy spurge reduces the livestock carrying capacity of pasture and rangeland by 20–50% with a subsequent loss of $35–45 million in U.S. beef and hay production (Bangsund and Leistritz, 1991).

Water

Many aquatic weeds complicate water utilization and management. Pervasive water weeds such as hydrilla (water thyme) in the southeastern United States and water hyacinth and common coontail globally can form large mats that interfere with transport of irrigation water and cooling water for hydroelectric plants. Such weeds may also interfere with recreational activities such as boating or fishing. Riparian weeds may also be heavy users of water, with a subsequent strain on water availability (Anderson, 1982). For example, salt cedar invasions in riparian areas of the southwestern United States are often blamed for depleting freshwater resources needed for irrigation or urban centers (although there is some controversy surrounding this point; Shafroth et al., 2005; Figure 1.4).

Environmental Damage

Some weeds may colonize a given habitat so successfully that native species may be eliminated altogether, with a subsequent decline in species diversity. Such weeds, usually nonnative for a given system, have proliferated to such a large extent that they are recognized by a separate classification; invasive or exotic weeds. Their introduction, commonly by human transport, results in widespread economic or environmental damage (Mooney and Hobbs, 2000). An example of the type of environmental damage related to the introduction of these invasive weeds is alteration of the fire cycle within native plant communities. These cycles are, in turn, accelerated by the introduction of fire-tolerant grass species such as cheatgrass, that can produce combustive vegetative with subsequent increases in the frequency, spread, and intensity of natural- or human-induced fire (Brooks et al., 2004). The conversion of shrub-dominated steppes of the western United States to a cheatgrass-dominated landscape during the twentieth century is a quintessential example of how invasive species introduction can significantly alter plant community composition over a large geographic area.

Cheatgrass is not the only species having dramatic effects on the landscape (Dukes and Mooney, 2004). Millions of acres of productive rangelands, forests, and riparian areas have been overrun by weedy invaders, with a subsequent loss of native flora. The proliferation

Figure 1.4 Salt cedar as a threat to water flow in the western United States. Accumulation of salt by this species may also alter soil salinity with negative effects on native plants. (John M. Randall, The Nature Conservancy.)

of aquatic weeds, in turn, can lead to greater eutrophication and significant disruptions in marine biology. It has been estimated that more than 200 million acres of natural habitats (primarily in the western United States) have already been lost to invasive, noxious weeds, with an ongoing loss of 3,000 acres a day (Westbrooks, 1998). The invasive plant species that are most harmful to native biodiversity are those that significantly change ecosystem processes (e.g., cheatgrass) to the detriment of native species.

Public Health

While the interaction of weeds and public health may seem somewhat obscure, weeds can directly affect public health through allergenic reactions, skin irritations, mechanical injury, and toxicology. Current estimates suggest there are approximately 30 million plant-based allergy sufferers in the United States. While many of these people are allergic to pollen of ornamental trees and grasses, a common weed, ragweed (*Ambrosia* spp.) is the principal source of plant-based allergens in the fall. A different kind of allergy, contact dermatitis or a skin-based allergic reaction, is also associated with weeds. Some examples include the milky sap in spurges or giant hogweed, stinging nettles, or urushiol contact with the poison ivy group. In the latter case, it is estimated that over 2 million people in the United States suffer annually from casual contact with poison ivy, poison oak, or poison sumac (Mohan et al., 2006).

Mechanical injury associated with weeds, from unwittingly removing a Canada thistle plant by hand to stepping barefoot on puncturevine seeds, can be a painful, although nonfatal

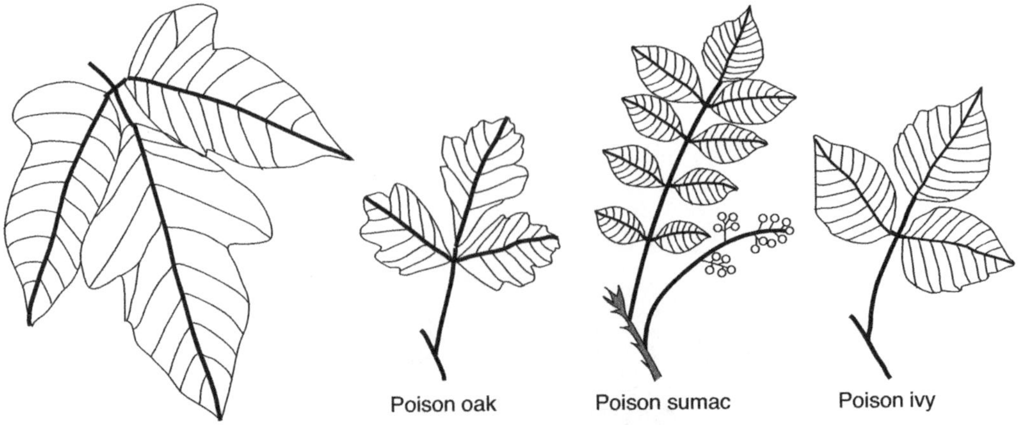

Figure 1.5 Poison ivy, a common weed of disturbed sites and contact dermatitis from poison ivy. (From USDA NRCS plants database.)

experience. Ingestion of poisonous weeds, however, can result in serious illness or death. There are over 700 plant species that are known to induce illness in humans or livestock. Some weed species such as poison hemlock, nightshade, or castor bean are so poisonous that even small quantities can be lethal. For example ricin, the poison contained within castor bean, has been linked to terrorist threats, in part, because ricin has a greater potency than cyanide (Figure 1.5).

There are other, more indirect means by which weeds impact human suffering. If weeds are unwanted plants, then plant-based sources of narcotics, from coca to poppy should be considered a significant threat to public health. Weeds may also indirectly alter air quality. Kudzu, for example, emits isoprene, a chemical precursor leading to tropospheric ozone formation (Sharkey and Loreto, 1993). Weeds can also affect the prevalence of human disease by providing food sources for disease vectors. Pollen, for example, can serve as a food source for mosquito larvae (Ye-Ebiyo et al., 2000), with potential interactions between mosquito populations and weed pollen.

Aesthetics

Aquatic weeds do not just interfere with water delivery, but detract from our enjoyment of the pleasures of nature, such as swimming in a clear lake. Avid gardeners spend many hours removing dandelions, chickweed, carpetweed, and garlic mustard because those plants take away from the colorful design the gardener wishes to create. Lawn enthusiasts also spend many hours (and many dollars) in cultivating a uniform green sward. And, as any home-owner can tell you, weeds can reduce property values. Yet, how can one quantify these impacts? Aesthetical deterioration of the environment is universally recognized, but remains ill-defined.

Figure 1.6 Examples of weed control: (1) Dr. John Teasdale, USDA-ARS weed scientist examines hairy vetch as potential cover crop in cultural weed control; (2) mechanical weed control through conventional tillage in soybean; (3) use of herbicides in no till field crop; (4) Dr. Rick Bennett, USDA-ARS Plant Pathologist, examines fungi that could be used for biological weed control. (Photos are USDA-ARS Image Gallery.)

Weed Management

Given the damage that weeds do to human activities, to the environment, and to public health, weed management remains, even today, the principal focus of weed science. Management, in turn, can be defined as those practices used to restrict the growth and spread of weeds. In general, these practices focus on three main areas; prevention, eradication, and control.

Prevention refers to those actions taken to inhibit the introduction and/or establishment of weed species into new areas. Eradication is defined as a set of measures taken to completely remove a weed species from an area. However, complete eradication is rarely achieved, in part, due to the high cost and effort involved in eliminating a weed species, particularly if that species has spread over a wide (>10 ha) area. Control is the more achievable outcome once a weed has become established, and refers to those methodologies that can reduce (but not eliminate) weed impacts to manageable levels. There are four recognized methods of weed control: (1) cultural, (2) mechanical, (3) chemical, and (4) biological (Figure 1.6).

Cultural

Cultural control recognizes and optimizes those practices common to good land and water management. At the farm level, these may include using weed-free crop seed, crop rotation to reduce the accumulation of high weed populations of certain species, and the use of smother crops or living mulches such as hairy vetch to control weed seed emergence (Teasdale and Mohler, 1993).

Mechanical

Mechanical control reflects some of the oldest known means to eliminate weed populations, including hand-pulling and hoeing. However, mechanical may also refer to flooding, a typical means of weed control in rice cultivation since rice can survive flooded conditions and many weeds cannot; the use of nonliving material to mulch the soil, usually as a means to prevent light interception from emergent weed seedlings; flaming or burning, utilizing controlled fire to burn young succulent weeds; and machine tillage, such as the use of chisel plows, weed knives, disc harrows, or some other tractor implement that breaks the soil and buries the weed.

Chemical

By far in the United States, and in most developing countries, chemical control of weeds through the selective application of herbicides is the most utilized practice. If applied properly and with respect to their utilization and safety, herbicides have been the most effective means to control weeds and to limit their negative impacts, particularly in crop production systems. For example, it has been estimated that in the United States, under best management practices, production losses associated with weeds averaged between 7 and 10%; however, without herbicides, those losses were between 35 and 38% (Bunce and Ziska, 2000; Table 1.2).

In 2002, approximately 78,000,000 ha of land in the United States was chemically treated to control weeds, grass, or brush; for comparison, 34,000,000 ha were treated chemically for disease, nematode, and insect control *combined*, for that same year (National Agricultural Statistical Service, www.nass.gov; Figure 1.7).

Because of their importance, many weed science texts are written specifically to address herbicide chemistry, focusing on physiological responses and a systematic evaluation of application methods that can increase uptake and overall efficacy of the herbicide. Such detail, however, is beyond the scope of this book, and we refer interested readers to excellent books by Duke (1985) and Monaco et al. (2002).

Table 1.2 Estimated production losses due to weeds for selected crops in the United States.

Crop	Best management practices ('000 Mt)	Best management practices without herbicides ('000 Mt)
Barley	261 (5)	1,045 (20)
Corn	19,967 (6)	99,837 (30)
Cotton	349 (8)	2,139 (49)
Potato	1,219 (6)	5,890 (29)
Rice	628 (7)	4,847 (54)
Sorghum	1,033 (8)	4,519 (35)
Soybean	5,838 (8)	27,731 (38)
Sugarcane	2,452 (9)	10.081 (37)
Wheat	3,382 (6)	11,273 (20)
Total	*35,129*	*167,362*

Production loss percentages (in parentheses) were calculated from data in Bridges (1992) and applied to 2007 production figures. Average crop losses due to weeds were 7% for best management practices and 35% for best management practices but without herbicide. Production values for sugarcane refer to bulk cane production.

2002 estimated annual agricultural use

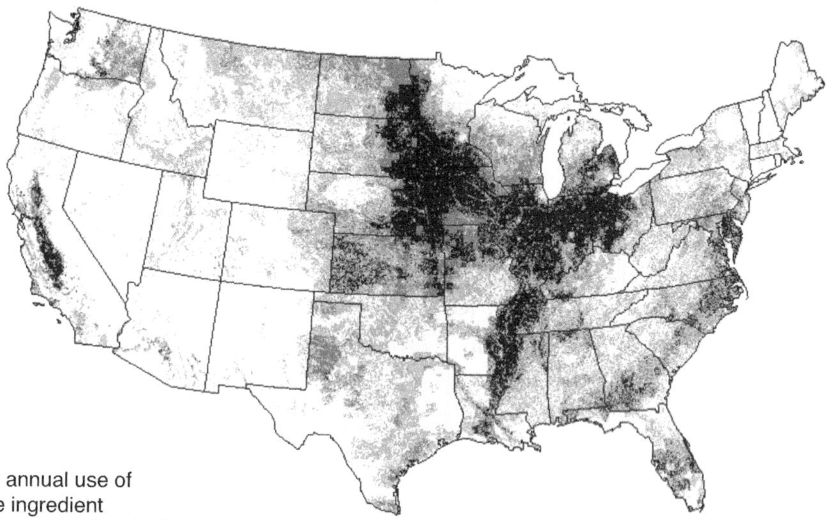

Average annual use of active ingredient (pounds per square mile of agricultural land in county)

- No estimated use
- 0.001–0.499
- 0.5–3.445
- 3.446–14.660
- 14.67–47.984

Crops	Total pounds applied	Percent national use
Soybeans	70,699,653	69.49
Corn	7,491,973	7.36
Cotton	7,063,492	0.94
Wheat for grain	3,509,151	3.45
Cropland in summer fallow	2,514,196	2.47
Citrus fruit	2,186,741	2.15
Sorghum	1,596,879	1.57
Rice	1,057,534	1.04

Figure 1.7 Estimated annual usage of glyphosate, the most frequently used herbicide in the United States. (From 2002, USGS pesticide use maps.)

In brief, herbicides are phytotoxic, usually organic compounds, divided into chemical families which include aliphatic and aromatic carboxylic herbicides, the substituted urea, carbamate, dintirophenol, dinitroaniline, *s*-triazine groups as well as the phosphonate/organic phosphorous compounds such as glyphosate. Nonorganic herbicides may include salts (e.g., copper sulfate) as well as acetic acid. Herbicides can be either selective or nonselective regarding the type of plant targeted, and may or may not travel to the site of action (i.e., systemic). Herbicides are either applied directly to emerged plants (postemergence) or directly to the soil (preplant or preemergence). Application of the herbicide may be as a spray in water or oil, or as dry granules. Added surfactants such as soap often increase absorption of the active ingredient by the plant.

Biological

Biological control is a strategy to employ a specific organism (parasite, predators, and pathogen) that once introduced will provide a measure of weed control, usually specific to the weed species. Biological control, if successful, can result in permanent weed control because the controlling organism is self-perpetuating. Another advantage is that the controlling organism, if benign, does not contribute to environmental pollution.

There are, however, a number of cautionary hurdles that must be overcome if biological control is to be effective. If the weed to be controlled is unwanted in one area, but valued in another, and both areas are within the same geographical region, biological control should not be used. Introduction of the biocontrol agent must be specific to the organism to be controlled and introduction of the controlling organism should not pose an additional environmental threat (e.g., cane toads introduced to Australia for biological control have become an enormous problem in themselves). The organism must tolerate similar environmental conditions as the weed species it is directed against, so it can survive in the same range. Overall, to produce an effective biocontrol measure requires money, time, and much effort on the part of the research community.

Integrated Pest Management

Integrated pest management (IPM) is not a single control measure per se, but an amalgamation of available means to monitor and control all pests (weeds, diseases, and insects) (Flint and van den Bosch, 1981). The driving force behind IPM was to develop a synergistic approach to monitoring and controlling pest populations. Overall, IPM is organized around six basic tenets:

1. Determination of acceptable pest levels, with the focus on control rather than eradication.
2. Use of cultural practices as previously described, in order to maintain healthy plants selected for a specific growing environment.
3. Monitoring of background pest levels. Regular observation is a fundamental principle of IPM. Visual inspection, traps, and other observational methods are employed to establish a history of potential pests for the region.
4. Mechanical control as a first line of defense. If a pest(s) reach unacceptable levels, mechanical options such as hand-pulling or tillage are utilized.
5. Chemical control as required. Emphasis on the use of chemicals as necessary and application of these chemicals at the most optimal time for successful weed control (e.g., greater efficacy of herbicides applied when weeds are in the seedling stage). The use of organic products in the field with limited environmental carryover whenever possible.
6. Development of long-term nonchemical solutions such as biological control.

Summary

It is impossible to summarize all that is known regarding weeds in a single chapter. It is hoped that the themes introduced here will provide an overview of the key aspects of weed biology, particularly how they are defined, characterized, and classified; the basic principles of weed ecology, focusing on the steps from seed germination to seed production; an overview of the negative impacts weeds have, both in human society and in nature; and, lastly but perhaps most importantly, methods of weed control by which these impacts can be minimized. In the following chapters, we will explore these themes in more detail, particularly in the context of anthropogenic climate change, the subject of our next chapter.

References

Anderson JE. 1982. Factors controlling transpiration and photosynthesis in *Tamarix chinensis* Lour. *Ecology* **63**:48–56.

Bangsund DA, Leistritz FL. 1991. *Economic Impacts of Leafy Spurge on Grazing Lands in the Northern Great Plains*. Agricultural Economic Report No. 275-S, North Dakota State University, Department of Agricultural Economics, Fargo, ND.

Bridges DC. 1992. *Crop Losses Due to Weeds in the United States*. Weed Science Society of America, Champaign, IL, 403 pp.

Brooks ML, D'Antonio CM, Richardson DM, Grace JB, Keeley JE, DiTomaso JM, Hobbs RJ, Pellant M, Pyke D. 2004. Effects of invasive alien plants on fire regimes. *Bioscience* **54**:677–688.

Bunce JA, Ziska LH. 2000. Crop ecosystem responses to climatic change: Crop/weed interactions. In: Reddy KR, Hodges HF (eds) *Climate Change and Global Crop Productivity*. CABI Publishing, New York, pp. 333–348.

Calvin M. 1949. The path of carbon in photosynthesis. VI. *J. Chem. Educ.* **26**:639–657.

Crafts AS. 1975. *Modern Weed Control*. University of California Press, Berkeley, CA, 440 pp.

Cudney DW, Jordan LS, Hall AE. 1991. Effect of wild oat (*Avena fatua*) infestations on light interception and growth rate of wheat (*Triticum aestivum*). *Weed Sci.* **39**:175–179.

Dauer J, Mortensen DA, Van Gessel M. 2007. Spatial and temporal dynamics governing long distance dispersal of *Conyza canadensis*. *J. Appl. Ecol.* **44**:105–114.

Diamond JM. 1997. *Guns Germs and Steel: The Fate of Human Societies*. WW Norton and Company, New York, 480 pp.

Donald CM. 1961. Competition for light in crops and pastures. In: *Mechanisms in Biological Competition, Soc. Exp. Biol. Symposia XV*. Academic Press, New York, pp. 282–313.

Duke SO. 1985. *Weed Physiology. Vol.2: Herbicide Physiology*. CRC Press, Boca Raton, FL, 283 pp.

Dukes JS, Mooney HA. 2004. Disruption of ecosystem processes in western North America by invasive species. *Rev. Chil. Hist. Nat.* **77**:411–437.

Eddington GE, Robbins WW. 1920. Irrigation water as a factor in the dissemination of weed seeds. *Colo Exp. Sta. Bull.* **253**:25.

Fairbrothers DE, Gray, JR (1972) *Microstegium vinineum* (Trin.) A. Camus (Gramineae) in the United States. *Bull. Torrey Bot. Club* **99**:97–100.

Flint ML, Van Den Bosch R. 1981. *Introduction to Integrated Pest Management*. Springer Press, New York, 256 pp.

Harper JL. 1977. *Population Biology of Plants*. Academic Press, New York, 892 pp.

Holm LG, Plucknett DL, Pancho JV, Herberger JP. 1977. *The World's Worst Weeds: Distribution and Biology*. East-West Center/University Press of Hawaii, Hawaii, 609 pp.

King LJ. 1966. *Weeds of the World, Biology and Control*. Interscience Publishers, New York, 270 pp.

Koch W. 1969. Influence of environmental factors on the seed phase of annual weeds, from the point of view of weed control. *Habilitations-schrift Landwehrshaft*, **50**:204–210.

Kropff MJ, Spitters CJT. 1991. A simple model of crop loss by weed competition from early observations on relative leaf area of the weeds. *Weed Res* **31**:97–105.

Labrada R. 2007. The need for improved weed management in rice. In: *Proceedings of the 20th Session of the International Rice Commission*, Bangkok, Thailand, July 23–26, FAO, Rome, pp. 310–324.

Miller JH. 2003. *Nonnative Invasive Plants of Southern Forests: A Field Guide for Identification and Control.* USDA, Forest Service, General Technical Report SRS-62, 93 pp.

Mohan JE, Ziska LH, Schlesinger WH, Thomas RB, Sicher RC, George K, Clark JS. 2006. Biomass and toxicity responses of poison ivy (*Toxicodendron radicans*) to elevated atmospheric CO_2. *Proc. Natl. Acad. Sci.* **103**:9086–9089.

Monaco TJ, Weller SC, Ashton FM. 2002. *Weed Science: Principles and Practices.* John Wiley & Sons, New York, 671 pp.

Mooney HA, Hobbs RJ. 2000. *Invasive Species in a Changing World.* Island Press, Washington, DC, 457 pp.

Novak SJ, Mack RN. 2001. Tracing plant introduction and spread: Genetic evidence from *Bromus tectorum* (Cheatgrass). *Bioscience* **51**:114–122.

Odum S. 1965. Germination of ancient seeds: Floristical observations and experiments with archaeologically dated soil samples. *Dan. Bot. Ark.* **24**:2.

Parsons WT. 1992. *Noxious Weeds of Australia.* Inkata Press, Melbourne, Australia.

Patterson DT. 1995. Effects of environmental stress on weed/crop interactions. *Weed Sci.* **43**:483–490.

Probert RJ. 1992. The role of temperature in germination ecophysiology. In: Fenner M (ed.) *Seed: The Ecology of Regeneration in Plant Communities.* C.A.B. International, Wallingford, UK, pp. 285–325.

Proctor VW. 1968. Long-distance dispersal of seeds by retention in digestive tract of birds. *Science* **160**:321–322.

Radosevich S, Holt J, Ghersa C. 1997. *Weed Ecology: Implications for Weed Management.* John Wiley & Sons, New York, 589 pp.

Scott HD, Geddes RD. 1979. Plant water stress of soybean (*Glycine max*) and common cocklebur (*Xanthium pennsylvanicum*): A comparison under field conditions. *Weed Sci.* **27**:285–289.

Shafroth PB, Cleverly JR, Dudley TL, Taylor JP, Van Riper C, Weeks EP, Stuart JN. 2005. Control of Tamarix in the Western United States: Implications for water salvage, wildlife use, and riparian restoration. *Environ. Manag.* **35**:231–246.

Sharkey TD, Loreto F. 1993. Water stress, temperature, and light effects on the capacity for isoprene emission and photosynthesis of kudzu leaves. *Oecology* **95**:328–333.

Skinner K, Smith L, Rice P. 2000. Using noxious weed lists to prioritize targets for developing weed management strategies. *Weed Sci.* **48**:640–644.

Sauer J, Struik G. 1964. A possible ecological relation between soil disturbance, light flash, and seed germination. *Ecology* **45**:884–886.

Stallings GP, Thill DC, Mallory-Smith CA, Lass LW. 1995. Plant movement and seed dispersal of Russian thistle (*Salsola iberica*). *Weed Sci.* **43**:63–69.

Teasdale JR, Mohler CL. 1993. Light transmittance, soil temperature, and soil moisture under residue of hairy vetch and rye. *Agron. J.* **85**:673–680.

Vengris J, Colby WG, Drake M. 1955. Plant nutrient competition between weeds and corn. *Agron. J.* **47**:213–216.

Westbrooks R. 1998. *Invasive Plants, Changing the Landscape of America: Fact Book.* Federal Interagency Committee for the Management of Noxious and Exotic Weeds (FICMNEW), Washington, DC, 109 pp.

Wilson RG. 1980. Dissemination of weed seeds by surface irrigation water in western Nebraska. *Weed Sci.* **28**:87–92.

Ye-Ebiyo Y, Pollack RJ, Spielman A. 2000. Enhanced development in nature of larval *Anopheles arabiensis* mosquitoes feeding on maize pollen. *Am. J. Trop. Med. Hyg.* **63**:90–93.

Zhang ZP. 2001. Weed management in rice in China. In: *Summary Presented at FAO Workshop on Echinochloa spp. Control*, Beijing, China, May 27, 2001.

Zimdahl RL. 1993. *Fundamentals of Weed Science*. Academic Press, New York, 454 pp.

2 Carbon Dioxide and Global Warming: The "Green" in the Greenhouse Effect

"Yes, it is real, sure is."

George W. Bush in an interview with the Politico answering the question, "For the record, is global warming real?", May 13, 2008

Introduction

It may be hard to believe, but the air you breathe in today, even in the most remote parts of the world, differs from the air that you breathed on the day you were born. Although you do not notice the difference, the plants around you, whether they are crops or weeds or native plants growing in the wild, certainly do. This chapter focuses on the atmosphere—primarily the change in the concentration of one gas, carbon dioxide—and the ways that the environment is responding. Put briefly, the climate is changing and plants are not growing quite like they used to. Why?

The Greenhouse Effect Made Simple!

For all practical purposes, Earth's climate depends on energy that arrives from the sun, and what happens to that energy once it reaches our atmosphere. If the sun's energy accumulated continuously on Earth, the planet would get warmer and warmer. In fact, our planet has reached, for now, a relatively stable temperature balance, where the energy coming in from the sun is balanced by energy leaving. Over tens and hundreds of thousands of years, this balance changes a bit, in what have been fairly regular cycles. For our purposes, though, these longer-term fluctuations are less relevant than what keeps the planet at its current temperature.

The sun is not the only object in space that sends energy our way. In fact, Earth receives energy from anything in space that is warmer than absolute zero (although on balance it loses energy to things cooler than itself). The warmer an object is, the more energy it loses to its surroundings via electromagnetic radiation, as per the Stefan–Boltzmann law. Because the sun is so hot, it gives off a tremendous amount of energy, and we benefit.

However, Earth also loses energy to space and that energy departs in its own spectrum of wavelengths. Because Earth is so much cooler (relative to the sun), less energy is lost, and the energy that departs has, on average, much longer (and thus less powerful) wavelengths. The departing energy falls into the infrared portion of the electromagnetic spectrum and is thus susceptible to absorption by greenhouse gases. The atmosphere's ability to trap infrared wavelengths—basically heat—depends on its composition. More of a given greenhouse gas will mean more energy absorption at the specific wavelengths absorbed by that gas. As different types of greenhouse gases enter the atmosphere, they increase energy absorption at a wider range of wavelengths. More energy absorption by the atmosphere means a warmer atmosphere, because less energy is lost to space. The system is sometimes described as a greenhouse, in that light can penetrate the greenhouse, but that the glass (i.e., atmosphere) prevents the long-wave infrared radiation (heat) from escaping. This is not, in itself, detrimental to biology; indeed, without this "greenhouse effect" the average temperature of the earth would be about $-18°C$!

What Is a Greenhouse Gas?

It turns out that many gases can absorb infrared wavelengths of light. The most common of these is water vapor. Others include carbon dioxide, methane, ozone, and many refrigerants. Some of these gases play a much larger role in changing the climate than others. The importance of a gas for climate change can depend on many factors such as how fast humans are releasing it, how much is already in the atmosphere, how long the gas stays in the atmosphere, and whether other gases already absorb the same wavelengths of energy.

Water vapor is the most abundant greenhouse gas, and humans have little influence on its atmospheric concentration. Water vapor varies dramatically from place to place; the warm air of the equatorial tropics can hold 10–20 times more water vapor than air at the poles.

When it comes to changing the climate, CO_2 is the most important greenhouse gas, primarily because we have changed its abundance in the atmosphere so dramatically—we have stuffed an additional 200+ billion metric tons of carbon, in the form of CO_2, into an atmosphere that used to hold less than 600 billion.

Several other greenhouse gases play important roles in climate change, including methane, nitrous oxide, and halocarbons (such as chlorofluorocarbons, or CFCs). Their relative importances are displayed in Table 2.1.

Where Do These Greenhouse Gases Come From?

Most of the major greenhouse gases were present in the atmosphere millions of years before humans appeared on Earth. However, people have dramatically changed the cycling of these gases. For example, every year, vast quantities of CO_2 are taken up by plants on land and by phytoplankton in the ocean. The carbon has been "fixed" by these organisms—integrated into carbohydrates and other compounds—for some period of time. As these organisms

Table 2.1 "Radiative forcing" values (RFs) for the most important greenhouse gases.

Common name	Radiative forcing (W m^{-2})
Carbon dioxide	1.56
Methane	0.59
Halocarbons	0.34
Nitrous oxide	0.15

These values describe how strongly the total emissions of these gases since 1750 were changing the climate in 2005. Data from Forster et al. (2007).

die and decompose, or as they use carbohydrates to power their metabolic processes, they release some of that carbon back to the atmosphere, again as CO_2. At the same time, CO_2 is continually dissolving into and back out of the ocean through physical processes. All these cycles are remarkably well balanced, leading to relatively stable CO_2 concentrations in the atmosphere over at least the last 800,000 years (Lüthi et al., 2008; Siegenthaler et al., 2005). However, during the last approximately 500 million years, some of the dying plants and phytoplankton have not fully decomposed, instead getting buried into a geological limbo of sorts, where time, pressure, and heat have incorporated some of their carbon into coal, oil, or natural gas (Dukes, 2003). This stored carbon largely powered the Industrial Revolution and the subsequent transformation of society, and its combustion has vented more than 300 billion tons of carbon into the air, principally as carbon dioxide (Canadell et al., 2007).

Agriculture and other activities have released another ~158 billion tons of carbon formerly stored in soils and plants (Canadell et al., 2007). Much of this was released from forests and forest soils as areas were converted to agricultural or grazing lands. These conversions, now primarily in tropical countries, still account for about 15–20% of all CO_2 releases worldwide (Canadell et al., 2007). Interestingly, the CO_2 released to the atmosphere by human actions has not all stayed there. We will discuss where this CO_2 has gone later in this chapter.

As with CO_2, the concentration of atmospheric methane (CH_4) has rapidly increased over the last 150 years, more than doubling during this time. Interestingly, the rate of growth has slowed dramatically since the early 1990s and the concentration may be approaching equilibrium with current emission rates (Forster et al., 2007). Methane is produced and consumed by widespread and common classes of microbes. The greatest rates of microbial production take place in wet (especially anaerobic) environments. Conversion of large areas to wet environments such as paddies for rice agriculture has increased methane production on the land surface. The growth of the livestock industry has also increased methane production because methane is generated in the digestive systems of ruminants. Methane is also released during the mining and distribution of fossil fuels, from landfills, and when biomass is burned.

Microbes in the soils and the ocean naturally emit nitrous oxide (N_2O) as they conduct nitrification and denitrification, but human activities have increased production of this gas by approximately 18% over preindustrial levels (Forster et al., 2007). Fertilized agricultural fields, especially when wet and when plants are taking up little or no nitrogen (such as when plants are still small or when crops have not yet been planted), provide ideal conditions for microbes to rapidly convert dissolved nitrogen into gaseous forms, including N_2O. Rapid denitrification also occurs when dissolved nitrogen runs off from these fields, and in manure management systems for livestock (Monteny et al., 2005; Stehfest and Bouwman, 2006). Other processes that create substantial amounts of N_2O include fossil fuel combustion, biomass burning, and some industrial processes such as nylon manufacturing (Denman et al., 2007).

Human actions are simultaneously destroying ozone in the stratosphere (the region from approximately 10 to 48 km in altitude) and stimulating its production in the troposphere (the atmosphere beneath the stratosphere). Different processes are operating at the different altitudes and the changes in concentration of this greenhouse gas have adverse health effects for plants (and humans) at each level. In the stratosphere, ozone molecules absorb ultraviolet (UV) radiation. As this ozone is destroyed through interactions with CFCs and other chemicals, more of the harmful radiation reaches the earth's surface, damaging nucleic acids and the photosynthetic apparatus. Consequently, UV damage may lead to decreased photosynthesis and plant growth (Caldwell et al., 2007). In humans, UV damage may lead to skin cancer. In the troposphere, releases of nitrogen oxides, volatile organic compounds, and other gases, often as products of fossil fuel combustion, increase the rate of ozone production. Ozone at Earth's surface enters leaves and stems and reacts with chemical constituents of plants, altering their biochemistry in ways that eventually reduce the growth of plant (Ashmore, 2005). In humans, ozone intensifies breathing ailments (asthma, emphysema, etc.). At all altitudes, ozone acts as a greenhouse gas.

Many synthetic halocarbons, including the CFCs that destroy ozone, are themselves powerful greenhouse gases, although they exist only in small concentrations in the atmosphere. Unlike the other greenhouse gases, most of the halocarbons are entirely manmade, for use as refrigerants, propellants, and other purposes. Production of many halocarbons has decreased or been halted following the passage of the Montreal Protocol, which was designed to phase out the manufacture of chemicals that led to depletion of stratospheric ozone.

Measuring Greenhouse Gases

The story of the realization that humans were changing the climate is well described by Weart (2003) and by Christianson (1999). In the 1950s, there was still some question about whether the carbon dioxide coming out of smokestacks and tailpipes would stay in the atmosphere. By comparing the isotopes, or kinds, of carbon in wood from old trees with those in new wood, the chemist Hans Seuss found that vegetation had taken up some of the newly released "fossil" carbon dioxide. It seemed quite possible—likely, even—that the vast majority of the carbon dioxide released by humans had dissolved in the ocean, leaving the atmosphere relatively unchanged. But by examining, among other things, the various isotopes of carbon present in the atmosphere, Seuss and his colleague Roger Revelle found evidence that the ocean was *not* rapidly taking up CO_2, and would only undo any changes in the atmosphere

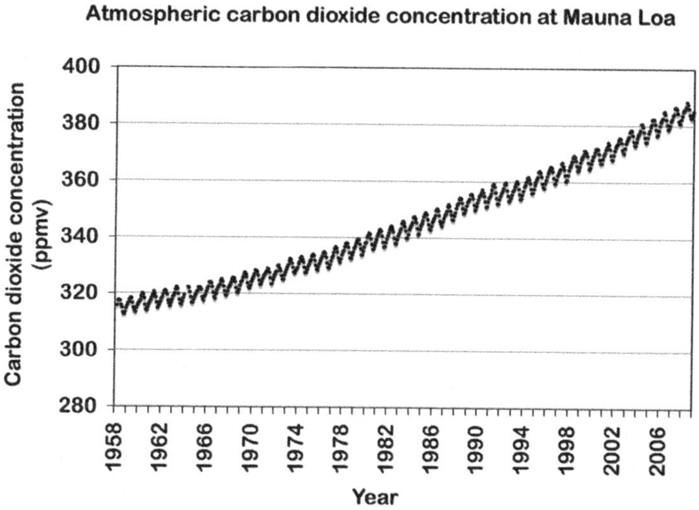

Figure 2.1 Atmospheric carbon dioxide concentration as measured at the monitoring station on Mauna Loa, Hawaii. This graph is also known as the "Keeling Curve." Collection of data continues today at this site. The annual fluctuations occur as a result of the seasons on Earth, and the greater land mass in the northern hemisphere. As plants take up carbon dioxide in the northern hemisphere during the spring and summer, atmospheric CO_2 concentrations decline. As plants in the northern hemisphere go dormant in fall and winter, CO_2 concentrations rise quickly. The accelerating long-term rise in the graph is attributable to fossil fuel burning and land use change. Throughout the record of monitoring on Mauna Loa, researchers have taken great care to factor out any gas samples contaminated with carbon dioxide from volcanic gases. Based on data from Keeling et al. (2009).

over many generations. Shortly after, Revelle hired a young atmospheric chemist, Charles Keeling, who might be able to show whether concentrations of the gas were increasing.

Keeling established the world's first long-term, highly sensitive CO_2 monitoring station in Hawaii, on the volcanic slopes of Mauna Loa. Data from this station now make up one of the most famous graphs in environmental science, known as the "Keeling Curve," which demonstrates how rapidly human activity has changed the atmosphere (Figure 2.1).

Initially, detecting an increase in atmospheric carbon dioxide concentrations was not a simple task; in the 1950s, no company made scientific instruments sensitive enough to detect CO_2 in the parts per million (ppm) range. Charles Keeling had taken it upon himself to better understand why Earth's atmosphere had so little of a gas that had such an important role in life. At CalTech, in 1955, he built the most sensitive instrument in existence to measure CO_2, and then carried out an obsessive spree of measurements. He started by carrying vacuum-containing glass spheres up to the roof every 4 hours, holding his breath, and then opening the flasks briefly, allowing them to fill with a sample of the region's air. This continued for days on end, during which he would leave his pregnant wife in the middle of the night to maintain his schedule of measurements. Soon afterward, he brought his patient wife and infant son into the mountains to camp in remote locations, as he again regularly filled flasks with air (here uncontaminated by the urban Pasadena environment). This obsession served him well.

After a couple of years, Revelle offered him a job at the Scripps Institution of Oceanography (where Seuss was also working). In his new position, Keeling was encouraged to follow his passion and to establish the first CO_2 monitoring stations.

It is interesting to note that Keeling's most valuable measurements almost did not get taken. Initial funding for Keeling's CO_2 station in Hawaii came from a year-long international campaign to measure geophysical processes, and Keeling's funding over the following 5 years was generously redirected from grants awarded to Revelle and Seuss at Scripps. Finally, though, the well ran dry. Government funding agencies were reluctant to fund a monitoring project to address a question that, at least until recently, was thought to be essentially irrelevant; after all, how could man's activities possibly change something so vast as Earth's atmosphere? Fortunately, Keeling finally convinced the U.S. National Science Foundation to fund his monitoring station in 1964, and the program resumed—leaving only a short gap in the record, and paving the way for a crucially important data set. If not for the establishment and maintenance of this monitoring project, the dramatic and ongoing impact of human activity on the atmosphere might not have been recognized for several more years.

Follow the Carbon (Where Will It Take Us?)

Today, many human actions increase atmospheric carbon dioxide concentrations. Most of the CO_2 released in industrial nations comes from combustion of coal, oil, and natural gas, which generates electricity and heat for buildings and which powers most transportation. A map of carbon dioxide releases from the United States shows how, at the local scale, much of the carbon dioxide is released from cities and roadways, as well as from more remote fossil–fuel burning power plants and industrial plants (Figure 2.2; Gurney et al., 2009). However, another source of carbon comes from the tropics, where conversion of forests to agricultural land results in massive carbon dioxide emissions as trees are burned or decompose. These "land use changes" account for approximately one-fifth of human-generated CO_2 releases (Canadell et al., 2007). All this carbon dioxide mixes relatively rapidly in the atmosphere.

The rate of carbon dioxide release has accelerated, almost continuously, since the beginning of the Industrial Revolution. The concentration had hovered at or below 280 ppm for at least the last 800,000 years, but as you read this, it has increased to about 390 ppm. This is

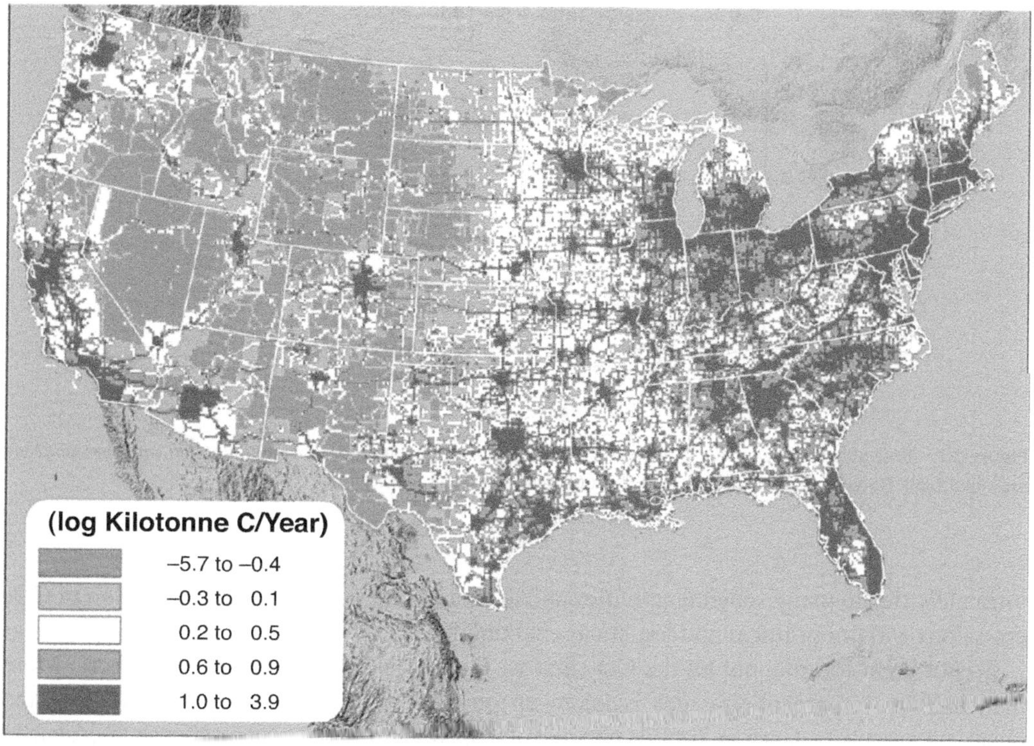

Figure 2.2 Spatial distribution of carbon dioxide release in the United States. Reprinted from Gurney et al. (2009). Available in color online at http://www.purdue.edu/eas/carbon/vulcan/plots.php.

probably the highest atmospheric CO_2 concentration of the last 20 million years (Canadell et al., 2007).

What drives this acceleration? Increasing human population, economic activity, and dependence on fossil fuel energy have acted together to accelerate CO_2 releases. In addition, population growth and economic pressures in developing countries have led to conversion of tropical forests to agricultural land. The "cooking" of limestone for cement manufacture also contributes a measurable fraction of human-generated CO_2, both because carbon dioxide is released from the limestone as it is heated and because the heating itself is an energy-intensive process. As population and world economies grow, demands for energy, forest conversion, and cement continue to rise (Figure 2.3; Boden et al., 2009; Houghton, 2008). How fast society releases CO_2 in the future will depend on several things, including the rate of human population growth and the rate at which economic growth fuels growth in energy consumption, but also the rate at which societies shift away from using fossil fuels and the rate at which energy efficiency increases. Each of these rates is related to many other factors in society, such as the rate of technological development, the dependence of nations on energy produced in other countries, and political realities and implications surrounding this dependence, the market prices of fossil fuels, and so on. Thus, CO_2 releases are

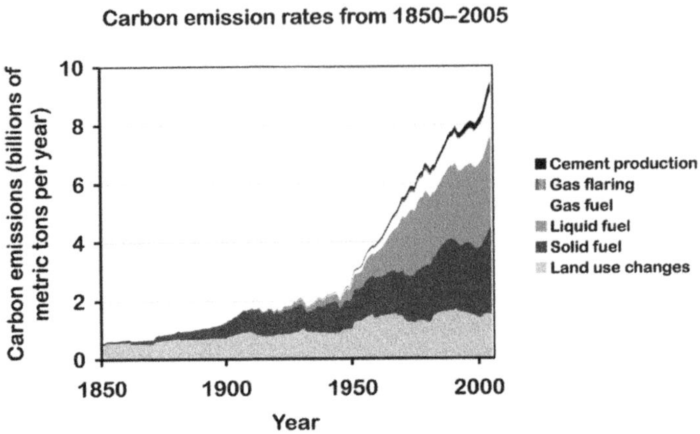

Figure 2.3 Rate of human-generated carbon emissions from fossil fuel burning, cement production, and changing land use since 1850. Based on data from Boden et al. (2009) and Houghton (2008).

intricately tied to many societal activities and decisions and this makes the regulation of CO_2 emissions a major political and logistical challenge.

To our great fortune, not all the CO_2 that we send into the atmosphere stays there. More than half of this greenhouse gas is taken up on land or at sea, thanks to biological and chemical processes. Carbon dioxide can dissolve in water and the oceans absorb one-quarter of the human-released CO_2, contributing among other things to their increasing acidity. Many marine organisms that form shells or other structures from carbonates, such as shellfish and corals, are threatened by these chemical changes (Doney et al., 2009). In contrast, carbon dioxide is also an essential ingredient for plant photosynthesis and growth (along with water, nutrients, and sunlight), and the increase in atmospheric CO_2 concentration, along with several other factors, has led to increased growth of plants on land. The additional carbon storage in these plants and in the soils beneath them accounts for another ~30% of the human-released CO_2 (Canadell et al., 2007). So, together, the oceans and ecosystems on land slow the growth in atmospheric CO_2 concentration to less than half the rate of CO_2 release.

The future composition of the atmosphere, then, depends not only on the rate of CO_2 release from human society, but also on the rate of CO_2 uptake by Earth's surface. Whether this rate of uptake will continue in the future is not certain—a variety of biological processes leading to uptake could conceivably saturate or slow in the coming decades. In recent years, uptake by Earth's surface has not kept pace with emissions, so the fraction of CO_2 that stays in the atmosphere has risen slightly, from about 40% in 1960 to approximately 45% today. This is primarily because the fraction of emitted CO_2 taken up by the oceans has declined (Canadell et al., 2007).

Given the complexity of the global carbon cycle, can we make any predictions about carbon dioxide concentrations in a future atmosphere? The most authoritative group when it comes to this general topic is the Intergovernmental Panel on Climate Change (IPCC). The IPCC is a body of scientists from around the world who periodically assess the latest knowledge about climate change. The consensus reports produced by this group are not

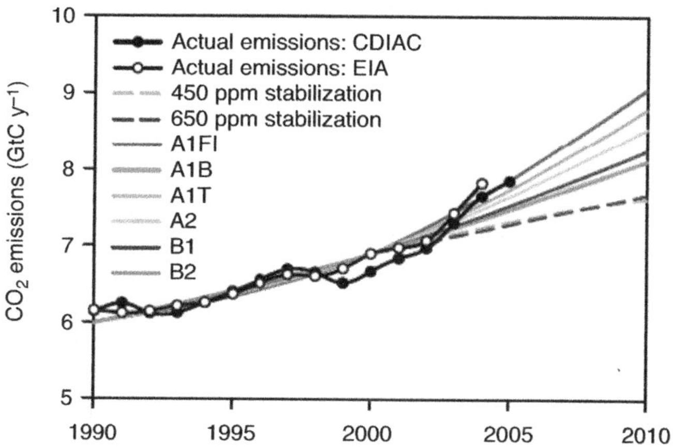

Figure 2.4 Observed CO_2 emissions according to two sources (black lines) compared with emissions scenarios used by the IPCC (solid gray lines) and emissions trajectories to achieve specified CO_2 concentration targets (dashed lines). From Raupach et al. (2007). Updated version available online in color at http://www.globalcarbonproject.org/carbonbudget/index.htm.

only vetted by many expert scientists at universities and research institutes, but also by governments and their scientists. These detailed, careful reports won the IPCC a share of the Nobel Peace Prize in 2007. In preparing the reports that examine how climate may change, the IPCC relies on a set of future scenarios that describe how society may change over the next 100 years, and how emissions of CO_2 and other greenhouse gases would change in each of those possible futures. Most of these scenarios are based on an assumption that, to one degree or another, technological progress will help society to wean itself off fossil fuels relatively quickly over the course of this century. The wide range of scenarios envisions that atmospheric CO_2 levels may rise to 463–623 ppm by 2050 and 478–1,099 ppm by 2100 (Carter et al., 2001). The highest emission scenario used by the IPCC is known as the A1FI scenario, with "FI" abbreviating "fossil intensive." Until the recent financial crisis, global human activity was producing as much or more carbon dioxide than was envisioned under this scenario, indicating that the IPCC's most recent worst-case scenarios for climate change could be somewhat optimistic (Figure 2.4; Raupach et al., 2007; Le Quéré et al., 2009).

Of course, carbon dioxide is not the only greenhouse gas affected by human actions. As mentioned previously, several other greenhouse gases (methane and nitrous oxide, along with ozone and CFCs) also have grown more concentrated in the atmosphere because of human actions (Figure 2.5). Water vapor, too, is becoming more concentrated in the atmosphere because of humans, but in this case the increase is a consequence of the addition of the other greenhouse gases. As these other gases warm the atmosphere, water evaporates more readily. Once in the atmosphere the additional water vapor acts as a potent greenhouse gas, which may cause more warming than even the greenhouse gases initially released by humans. More warming, in turn, leads to more water vapor. This phenomenon is known as a positive feedback. The buildup of greenhouse gases is causing the atmosphere to trap heat more effectively, warming the surface of the planet. But how much more will it warm, where will it warm the most, and how do we know?

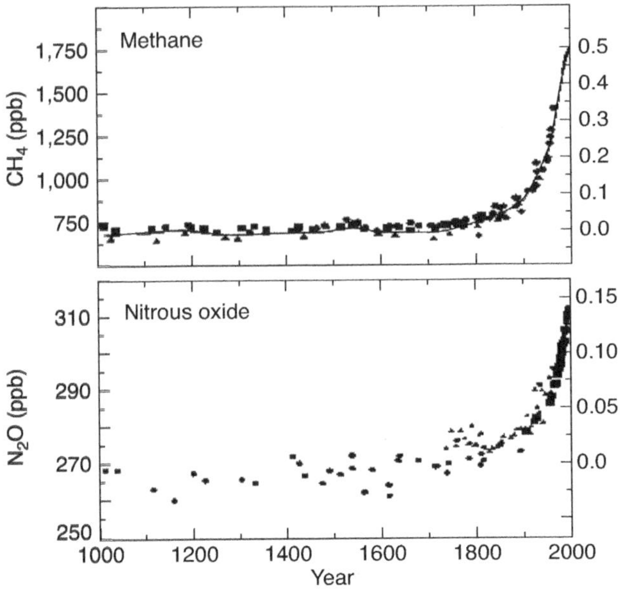

Figure 2.5 Graph of increasing CH_4 and N_2O concentrations. Atmospheric concentration is shown on the left axis, and radiative forcing, or the relative influence of each gas on Earth's energy balance, is shown on the right axis (in W m^{-2}). Figure from IPCC (2001).

Because the basic properties of greenhouse gases are known, we can estimate how well a given atmosphere would insulate Earth's surface from space. Clearly, increasing the concentration of greenhouse gases should cause the surface to warm. How much it will warm is a much more complicated question because of the many feedback loops in Earth's climate such as the positive feedback from water vapor. Another positive feedback loop may be related to albedo or the amount of energy reflected back into space. For example, warming will mean that less of the planet is covered by ice and snow. In turn, less of the sun's energy will be reflected back to space, and instead water, plants, and soil will capture that energy, warming the planet further. A slightly warmer planet will mean that trees can grow at higher elevations and latitudes, making portions of Earth's surface darker still. Some other potentially important feedbacks (positive and negative) such as the consequences of increased cloudiness are less well understood.

Energy absorbed by Earth's surface gets redistributed on a vast scale by ocean currents. Winds and storm systems also rearrange large amounts of energy. Our knowledge about the workings of the climate system, the properties of greenhouse gases, and the workings of climate feedbacks are all necessary to predict how our actions will affect Earth's climate.

Looking into the Crystal Ball: Climate Models and Projections

The best attempts by climate scientists to understand how differential changes in surface temperatures might alter climate are based on complex mathematical representations of the physical world: models that simulate the workings of the climate system on high-performance computers. When the IPCC issues reports containing "climate projections," these reports are their best attempts to summarize what has been seen in these crystal balls. (Climate modelers do not use the term "prediction" because they do not know what the actual rate of emissions will be; each projection is essentially a prediction for the future *if* a certain set of conditions comes to pass.)

Researchers in many different countries have created dozens of climate models over the years, each with its own particular design and strategy for representing the climate system and its various feedbacks. As more of the Earth system's processes have been incorporated into these models, the models have become more accurate at simulating past climates. The IPCC summarizes results from the broad spectrum of climate models, across a variety of future emission scenarios. While these models typically differ in their specific projections, the general forecasts for warming and, in many areas of the world, precipitation, are remarkably consistent (Solomon et al., 2007).

Under all the emission scenarios, the models foresee warming over the rest of this century. The rate of warming depends on the emission scenario, although models project relatively similar amounts of warming in the next few decades across diverse scenarios. However, later in the century, different emission rates make major differences in projected climates.

There are some general consistencies among models. First, warming will be greatest over the land, as opposed to the ocean. This means that the estimates of "average" warming for the globe reported in the media are typically lower than people and plants would actually experience. Second, warming will be greatest at the poles, most dramatically at high northern latitudes. Such projections are consistent with many of the recent data regarding polar warming, increasing desertification, warmer winters, etc. Although future emission rates will determine the intensity of warming, the spatial pattern of warming is consistent across different scenarios (Figure 2.6; Meehl et al., 2007).

Climate models project shifts in precipitation patterns as well as temperature. In general, precipitation is projected to increase at high latitudes, particularly in the winter. Many regions of the world that are currently relatively dry are projected to become drier. These regions, including the Mediterranean, southern Africa, and western Mexico, could be subject to increased desertification (Figure 2.7). Globally, precipitation is projected to come in fewer events per year; basically, storms will be more intense and arrive less often. In most regions, this is bad news from the perspective of a plant; effectively, less of the precipitation that falls will be available because more will run off the land during large storms. As the time between storms increases, short droughts may become more frequent.

The climate models considered by the IPCC have a fairly low spatial resolution, largely because the models demand so much computing power to run. Unfortunately, this coarse scale prevents the models from accurately capturing effects of many local features such as mountain ranges. Two techniques are used most often to create climate projections with local detail. Regional climate models have been created to simulate climate in particular regions, typically at the subcontinental scale. Statistical "downscaling," working at similar scales, interprets the output of the global climate models in the context of the typical local climate

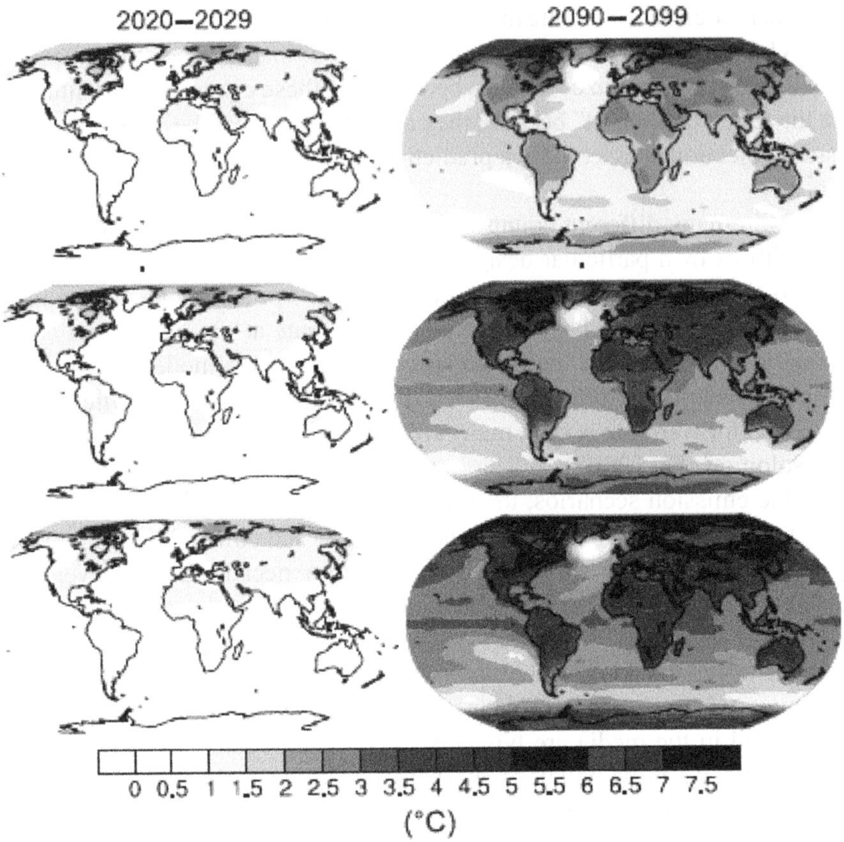

Figure 2.6 Projected warming of Earth's surface (relative to temperatures in 1980–1999) under scenarios of lower, intermediate, and higher greenhouse gas releases. Until recently, the actual emission rate exceeded that used in the highest release scenario. Note the consistent spatial variation in projected warming. Figure from Meehl et al. (2007). Available in color online at http://www.ipcc.ch/publications_and_data/ar4/wg1/en/tssts-5-2.html

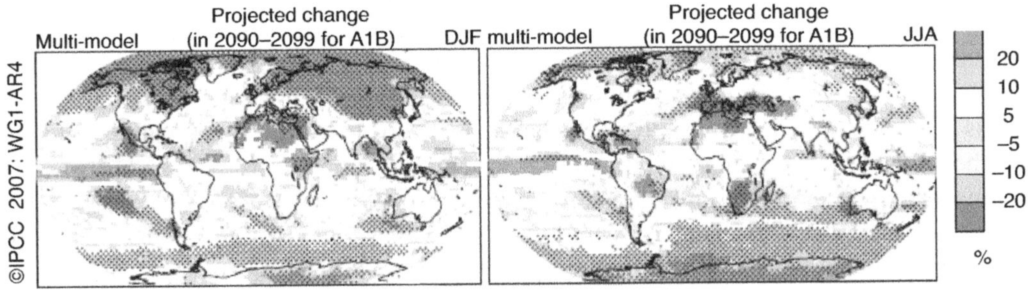

Figure 2.7 Projected seasonal precipitation changes (relative to temperatures in 1980–1999) under a scenario of intermediate greenhouse gas releases. The figures show the mean responses across *several* climate models. Areas with black dots indicate that more than 90% of the models used agree on the nature of the change. The left panel shows changes in December through February and the right panel shows changes in June through August. Figure from Solomon et al. (2007). Available in color online at http://www.ipcc.ch/publications_and_data/ar4/wg1/en/tssts-5-3.html

relationships seen within a region. The products of these regional approaches are particularly useful when contemplating how particular ecosystems may change in the future and they provide opportunities for people in the region to consider how climate change will affect them.

A Plant's "Eye" View of the Future

How will plants respond to these changing conditions? This is not a trivial question, since plants are the basis of the global food chain and any change in their biology will impact all living systems. For obvious reasons, therefore, questions regarding plant function have been the subject of thousands of scientific papers and talks. But as many researchers have determined, assessing the impact of climate change on plant biology is difficult, particularly on a large scale. First of all, people are changing many aspects of the environment simultaneously. The gas that plants use as a building block, carbon dioxide, has become nearly 40% more abundant in just over a century. Imagine how we would respond if oxygen were 40% more abundant! (Well, we might have massive problems with wildfires, but certainly we would all be better athletes!) We are moving plants and their diseases, insect pests, and pollinators from continent to continent around the world, so that plant species are now interacting with a different community than that in which they most recently evolved. We are appropriating habitat and altering it to serve our needs, causing the effective sizes of many plant populations to shrink (with the notable exceptions of crops and many weed species), and reducing the amount of available habitat in which seeds might germinate and survive. On the other hand, we are regularly disturbing vast areas of land along roadsides, train tracks, in cities, and in other areas associated with human activity, killing or cutting the native vegetation in these areas and creating conditions in which fast-growing weeds can thrive. We are fertilizing the world, on purpose as well as accidentally, via the nitrogen-containing constituents of exhaust. We are spraying vast tracts of land with a variety of herbicides to benefit crops and other desired species. And, on top of all this, we are changing the climate. These changes are not all happening at the same rate in the same locations, but no plants are escaping the changing atmosphere. We may develop decent predictions about how plants will respond to one of these variables, but how might a second variable affect that response?

It is also difficult to predict how plants as a group will respond to future conditions because there are so many different species and vegetation types, some of which are particularly difficult to study. Imagine, for example, trying to characterize how tropical rainforest will respond to one of these perturbations, even in a single region. This would involve trying to understand the aggregate response of hundreds of different tree species (not to mention the other plants), many of which live for hundreds of years. How should one choose which species to study, and which life stages? Even then, how would we know whether the plant's response might change a relationship with a herbivore or a pollinator, and whether that changing relationship might have even more important consequences for the forest than the plant response itself? And still, all this ignores the practical challenges of working in such an environment.

Despite these and other challenges, researchers have made progress. Much is understood about the basic physiological responses of plants to increased levels of atmospheric carbon dioxide, to warming, and to changes in precipitation.

It was suspected early on that these physiological responses would depend on the type of photosynthetic machinery used by a plant. A variety of different photosynthetic pathways are found in the plant kingdom, which can be loosely categorized as C_3, C_4, and CAM photosynthesis—terms derived from the initial products of carbon fixation in each pathway. C_4 species like sugarcane and corn are mostly found in hot, dry environments, where their physiology gives them a few competitive advantages. For one thing, they use water efficiently—for each milliliter of water they lose through their leaves, they can grow more than plants using the C_3 pathway. Second, at high temperatures, they typically carry out photosynthesis faster than C_3 plants. Our changing atmosphere is identifying (or creating) a drawback of their photosynthetic system, though: It is less responsive to high CO_2 concentrations than that of C_3 species. From these basic properties, we might expect warming to favor C_4 species, and increasing CO_2 to favor C_3s. So, how will these different types of species fare, in a warmer, more CO_2-rich world? Which species will be more competitive in the future? We will discuss this more in the next chapter.

More will determine a species' future success than its photosynthetic pathway, of course. Many plants respond to environmental changes by altering their allocation to different tissue types and changing their internal chemistry. For instance, plants grown in elevated CO_2 typically have lower concentrations of nitrogen-containing proteins in their leaves. This has consequences for the species that eat them, as these animals (or people) will get less protein per mouthful in the future. Other chemical changes, such as altered levels of plant toxins, could also change plant–animal interactions. Some of these chemical changes may affect the way that weed species respond to management techniques such as the application of herbicides. Changes in allocation to roots and shoots could also affect management techniques. We will explore the possible consequences in the next chapter.

Summary

Human actions have injected vast quantities of greenhouse gases into the atmosphere over the past approximately 150 years. These include CO_2, released primarily from combustion of fossil fuels, methane, and nitrous oxide, released by agriculture and industry, and other gases. Because they can absorb infrared radiation emitted from Earth's surface, these gases cause Earth's atmosphere to warm up more than it previously could have. This, in turn, keeps Earth's surface warmer. Climate models suggest significant warming will occur by 2100, particularly over land. This will be accompanied by changes in precipitation patterns. Plants will respond directly to the increase in CO_2 concentration, as well as to climate change and other ongoing environmental changes. In the next two chapters, we will begin to address in more detail the reported and potential changes in plant function resulting from climate change, focusing in particular on weed biology.

References

Ashmore MR. 2005. Assessing the future global impacts of ozone on vegetation. *Plant Cell Environ.* **28**:949–964.

Boden TA, Marland G, Andres RJ. 2009. *Global, Regional, and National Fossil–Fuel CO_2 Emissions.* Carbon Dioxide Information Analysis Center, Oak Ridge National Laboratory, U.S. Department of Energy, Oak Ridge, TN.

Caldwell MM, Bornman JF, Ballaré CL, Flint SD, Kulandaivelu G. 2007. Terrestrial ecosystems, increased solar ultraviolet radiation, and interactions with other climate change factors. *Photochem. Photobiol. Sci.* **6**:252–266.

Canadell JG, Le Quéré C, Raupach MR, Field CB, Buttenhuis ET, Ciais P, Conway TJ, Gillett NP, Houghton RA, Marland G. 2007. Contributions to accelerating atmospheric CO_2 growth from economic activity, carbon intensity, and efficiency of natural sinks. *Proc. Natl. Acad. Sci. U. S. A.* **104**:18866–18870.

Carter TR, La Rovere EL, Jones RN, Leemans R, Mearns LO, Nakicenovic N, Pittock AB, Semenov SM, Skea J. 2001. Developing and applying scenarios. In: McCarthy JJ, Canziani OF, Leary NA, Dokken DJ, White KS (eds) *Climate Change 2001: Impacts, Adaptation, and Vulnerability.* Cambridge University Press, Cambridge, UK, pp. 145–190.

Christianson GE. 1999. *Greenhouse: The 200-Year Story of Global Warming.* Walker and Company, New York.

Denman KL, Brasseur G, Chidthaisong A, Ciais P, Cox PM, Dickinson RE, Hauglustaine D, Heinze C, Holland E, Jacob D, Lohmann U, Ramachandran S, Dias PLDS, Wofsy SC, Zhang X. 2007. Couplings between changes in the climate system and biogeochemistry. In: Solomon S, Qin D, Manning M, Chen Z, Marquis M, Averyt KB, Tignor M, Miller HL (eds) *Climate Change 2007: The Physical Science Basis. Contribution of Working Group I to the Fourth Assessment Report of the Intergovernmental Panel on Climate Change.* Cambridge University Press, Cambridge, UK, pp. 499–587.

Doney SC, Fabry VJ, Feely RA, Kleypas JA. 2009. Ocean acidification: the other CO_2 problem. *Ann. Rev. Marine Sci.* **1**:169–192.

Dukes JS. 2003. Burning buried sunshine: human consumption of ancient solar energy. *Clim. Change* **61**:31–44.

Forster P, Ramaswamy V, Artaxo P, Berntsen T, Betts R, Fahey DW, Haywood J, Lean J, Lowe DC, Myhre G, Nganga J, Prinn R, Raga G, Schulz M, Dorland RV. 2007. Changes in atmospheric constituents and in radiative forcing. In: Solomon S, Qin D, Manning M, Chen Z, Marquis M, Averyt KB, Tignor M, Miller HL (eds) *Climate Change 2007: The Physical Science Basis. Contribution of Working Group I to the Fourth Assessment Report of the Intergovernmental Panel on Climate Change.* Cambridge University Press, Cambridge, UK, pp. 129–234.

Gurney KR, Mendoza DL, Zhou Y, Fischer ML, Miller CC, Geethakumar S, de la Rue du Can S. 2009. High resolution fossil fuel combustion CO_2 emission fluxes for the United States. *Environ. Sci. Tech.* **43**:5535–5541.

Houghton RA. 2008. Carbon flux to the atmosphere from land-use changes: 1850–2005. In: *Trends: A Compendium of Data on Global Change.* Carbon Dioxide Information Analysis Center, Oak Ridge National Laboratory, U.S. Department of Energy, Oak Ridge, TN.

Keeling RF, Piper SC, Bollenbacher AF, Walker JS. 2009. Atmospheric CO_2 records from sites in the SIO air sampling network. In: *Trends: A Compendium of Data on Global Change.* Carbon Dioxide Information Analysis Center, Oak Ridge National Laboratory. U.S. Department of Energy, Oak Ridge, TN.

Le Quéré C, Raupach MR, Canadell JG, Marland G, Bopp L, Ciais P, Conway TJ, Doney SC, Feely RA, Foster P, Friedlingstein P, Gurney K, Houghton RA, House JI, Huntingford C, Levy PE, Lomas MR, Majkut J, Metzl N, Ometto JP, Peters GP, Prentice IC, Randerson JT, Running SW, Sarmiento JL, Schuster U, Sitch S, Takahashi T, Viovy N, van der Werf GR, Woodward FI. 2009. Trends in the sources and sinks of carbon dioxide. *Nature Geoscience* **2**:831–836.

Lüthi D, Floch ML, Bereiter B, Blunier T, Barnola J-M, Siegenthaler U, Raynaud D, Jouzel J, Fischer H, Kawamura K, Stocker TF. 2008. High-resolution carbon dioxide concentration record 650,000–800,000 years before present. *Nature* **453**:379–382.

Meehl GA, Stocker TF, Collins WD, Friedlingstein P, Gaye AT, Gregory JM, Kitoh A, Knutti R, Murphy JM, Noda A, Raper SCB, Watterson IG, Weaver AJ, Zhao Z-C. 2007. Global climate projections. In: Solomon S, Qin D, Manning M, Chen Z, Marquis M, Averyt KB, Tignor M, Miller HL (eds) *Climate Change 2007: The Physical Science Basis. Contribution of Working Group I to the Fourth Assessment Report of the Intergovernmental Panel on Climate Change*. Cambridge University Press, Cambridge, UK, pp. 747–845.

Monteny G-J, Bannink A, Chadwick D. 2005. Greenhouse gas abatement strategies for animal husbandry. *Agric. Ecosyst. Environ.* **112**:163–170.

Raupach MR, Marland G, Ciais P, Quéré CL, Canadell JG, Klepper G, Field CB. 2007. Global and regional drivers of accelerating CO_2 emissions. *Proc. Natl. Acad. Sci. U. S. A.* **104**:18866–18870.

Siegenthaler U, Stocker TF, Monnin E, Lüthi D, Schwander J, Stauffer B, Raynaud D, Barnola J-M, Fischer H, Masson-Delmotte V, Jouzel J. 2005. Stable carbon cycle–climate relationship during the late Pleistocene. *Science* **310**:1313–1317.

Solomon S, Qin D, Manning M, Alley RB, Berntsen T, Bindoff NL, Chen Z, Chidthaisong A, Gregory JM, Hegerl GC, Heimann M, Hewitson B, Hoskins BJ, Joos F, Jouzel J, Kattsov V, Lohmann U, Matsuno T, Molina M, Nicholls N, Overpeck J, Raga G, Ramaswamy V, Ren J, Rusticucci M, Somerville R, Stocker TF, Whetton P, Wood RA, Wratt D. 2007. Technical summary. In: Solomon S, Qin D, Manning M, Chen Z, Marquis M, Averyt KB, Tignor M, Miller HL (eds) *Climate Change 2007: The Physical Science Basis. Contribution of Working Group I to the Fourth Assessment Report of the Intergovernmental Panel on Climate Change*. Cambridge University Press, Cambridge, UK, pp. 19–91.

Stehfest E, Bouwman L. 2006. N_2O and NO emission from agricultural fields and soils under natural vegetation: summarizing available measurement data and modeling of global annual emissions. *Nutr. Cycling Agroecosyst.* **74**:207–228.

Weart S. 2003. *The Discovery of Global Warming*. Harvard University Press, Cambridge, MA.

3 An Evaluation of the Impact of Rising Carbon Dioxide and Climatic Change on Weed Biology: From the Cell to the Plant

"Though I do not believe that a plant will spring up where no seed has been, I have great faith in a seed. Convince me that you have a seed there, and I am prepared to expect wonders."

Henry David Thoreau

Introduction

We have seen from the previous chapter that increases in carbon dioxide concentration CO_2 and the subsequent consequences for climate, particularly surface temperature and hydrology, are associated with human activities, most notably the demand for energy (fossil fuel consumption) and land (deforestation). Given these projected environmental changes, what are the likely impacts as they relate to our understanding of the basic biology of weeds?

To answer this question, one needs to understand basic plant biology, how rising CO_2 and climate will alter plant biology on a number of different functional levels, and how those levels interact with each other. To appreciate the complexity of plants, we must keep in mind that biological processes at the cellular level (e.g., photosynthesis) happen quickly and at very small scales, whereas those at the ecosystem level take longer and occur over larger areas (e.g., plant-to-plant competition for resources, seed production) (Figure 3.1).

In this chapter, we examine these biological processes in detail and discuss how rising CO_2 and the combination of rising CO_2 and climate (precipitation and temperature) are likely to impact weed biology. We consider different levels of functional organization from the cellular level to the whole plant, using the construct as shown in Figure 3.1. The following chapter will "scale-up" from the response of individuals to examine responses of plant–plant interactions and biogeographical distribution.

This approach recognizes the certainty of continuing increases in global levels of atmospheric CO_2, but also acknowledges that the extent to which the increase in CO_2 and other anthropogenic greenhouse gases alter surface temperature and water availability is likely to vary temporally and geographically (Solomon et al., 2007).

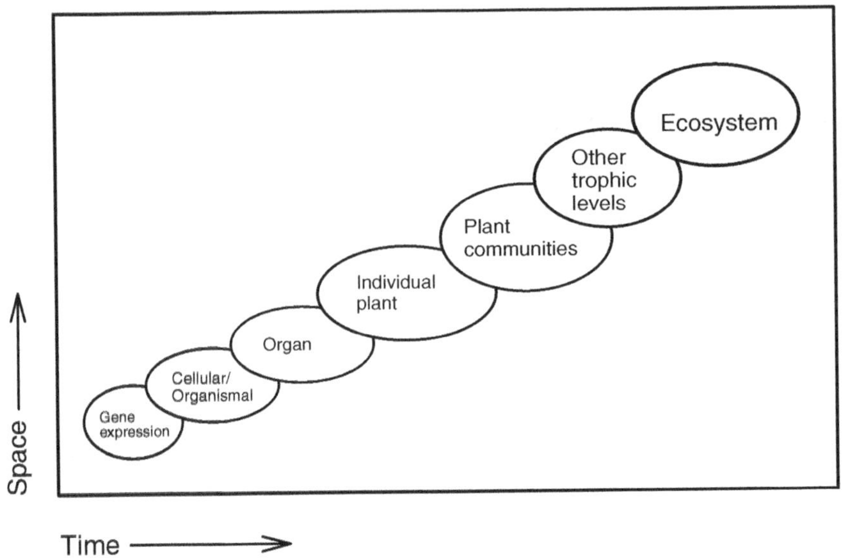

Figure 3.1 Theoretical construct of biological processes in space and time.

Gene Expression and Carbon Dioxide

The influence of projected future increases in CO_2 concentration on gene expression, particularly as it relates to photosynthetic regulation of the small subunit of Ribulose-1,5-bisphosphate carboxylase (Rubisco) have been examined in a number of studies (Cheng et al., 1998; Makino et al., 2000; Moore et al., 1999). In these instances, genetic regulation is thought to be mediated by increased sugar levels resulting from exposure to future CO_2 concentrations (e.g., Cheng et al., 1998); others however, have argued that any CO_2-induced decline in photosynthetic gene transcripts is due to a temporal shift in leaf ontogeny (Ludewig and Sonnewald, 2000).

Although a number of reviews have examined how carbohydrate accumulation may modify genetic regulation of both photosynthetic and nonphotosynthetic genes (Koch, 1996; Sheen, 1994); the specific function of rising CO_2 on the subsequent change in carbohydrate signaling has not been fully examined. Unpublished data for corn grown in Soil, Plant, Atmosphere Research (SPAR) units indicated that approximately 5% of the genome responded to elevated CO_2 (750 ppm) (Soo-Hyung Kim, personal communication). However, it was difficult to link this result to physiological responses partly because many of these genes encoded unknown or putative functional proteins. *Arabidopsis thaliana* is not a significant weed; however, when the species was grown under elevated CO_2, transcription levels of some nonphotosynthetic genes for growth, development, and stress were increased (Bae and Sicher, 2004).

Changes in gene expression may provide crucial insight into specific mechanisms or cellular systems that may be regulated by changes in atmospheric CO_2. The mechanistic basis for such changes, whether they involve carbohydrate accumulation or accelerated ontogeny is unclear. In addition, it is not clear if the mechanistic basis differs between weeds and crops. Consequently, analyses of transcript profiles from microarray experiments, particularly from crops and weeds of the same genus, or wild and cultivated types of the same species grown over a range of carbon dioxide values and/or temperatures, may be of particular benefit in understanding potential environmental fitness for managed crop and weed species in the future.

Photosynthesis: Cell and Leaf

Plants are able to grow by synthesizing what they need from their environment. Essential to this process is photosynthesis, the conversion of light to chemical energy. The raw materials needed for photosynthesis are carbon dioxide, water, light, and nutrients, particularly nitrogen and phosphorous. Temperature is also essential to the process in that it must be within a set of thermal limits for photochemistry to succeed.

Land plants evolved approximately 450 million years ago, appearing at a time when atmospheric concentrations of carbon dioxide were between 1,000 and 1,200 ppm. However, since the early Miocene (about 24 million years BP), CO_2 concentrations appear to have declined below 500 ppm (Pearson and Palmer, 2000), and evidence from ice core data indicates a CO_2 fluctuating between 200 and 300 ppm for at least the last 800,000 years (Lüthi et al., 2008; Siegenthaler et al., 2005). Because CO_2 represents the sole source of carbon for photosynthesis, and because atmospheric levels have been low for the recent geological past, the current anthropogenic increases in CO_2 (+22% since 1960), represents an upsurge of an essential resource needed for plant growth.

For the C_3 photosynthetic pathway (labeled "C_3" because the first product is a three-carbon compound, phosphoglycerate, 3-PGA), assimilation of carbon dioxide is dependent on the concentration and kinetic properties of the enzyme Ribulose-1,5-bisphosphate carboxylase/oxygenase (Rubisco). Because the enzyme is both a carboxylase and oxygenase, rising CO_2 stimulates photosynthetic rates by raising the CO_2 gradient from air to leaf, and by reducing the loss of CO_2 by suppressing photorespiration. (Photorespiration occurs when Rubisco fixes oxygen instead of carbon dioxide resulting in the net release of CO_2.) The increase in photosynthesis as a result of recent and projected increases in CO_2 is one of the most-researched aspects of rising CO_2 concentration. Of the 250,000+ plant species, approximately 95% obtain CO_2 solely from the C_3 photosynthetic pathway, and there are, literally, hundreds of studies showing that both recent and projected increases in atmospheric carbon dioxide can significantly stimulate photosynthesis (see Kimball, 1983; Kimball et al., 1993; Poorter and Navas, 2003, for reviews examining the response to future CO_2 concentrations; Sage, 1995, for a review of the response to recent CO_2 increases).

The advent of C_4 photosynthesis during the Miocene may well have been a photosynthetic modification that evolved in response to declining CO_2 and warmer climates that exacerbated CO_2 loss through photorespiration in C_3 plants. The C_4 photosynthetic pathway (labeled "C_4" because the first product is a four-carbon compound, oxaloacetate), adds an additional

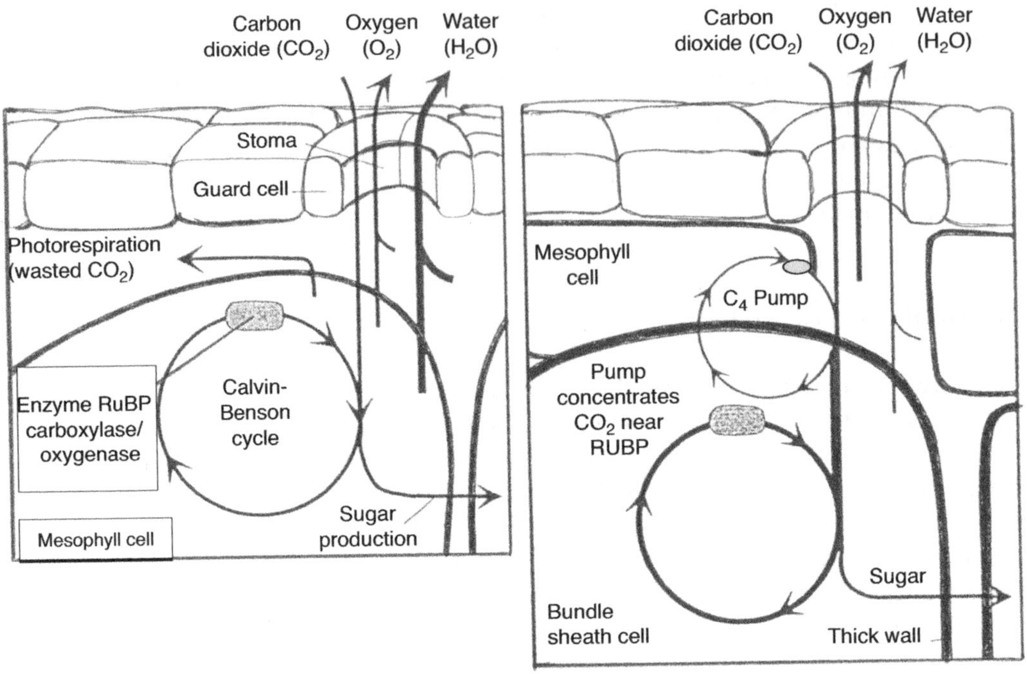

Figure 3.2 Schematic of C_3 and C_4 photosynthesis. Note that the presence of phosphoenolpyruvate (PEP) carboxylase acts to concentrate CO_2 around Rubisco and minimize CO_2 lost through photosynthetic carbon oxidation.

mechanism similar to a "supercharger" in a combustion engine, whereby CO_2 is concentrated around Rubisco in order to reduce or eliminate the oxygenase component of the enzyme. This is done using a combination of anatomical compartmentalization within the leaf and a different enzyme to initially fix CO_2. Rubisco is isolated within bundle sheath cells surrounding the vascular bundles, allowing a more rapid export of photosynthetic products from the leaf. Mesophyll cells can then pump CO_2 into these cells increasing the relative proportion of CO_2 to O_2, thereby reducing the amount of carbon lost via photorespiration. The pump is powered by sunlight, and operates using the enzyme phosphoenolpyruvate (PEP) carboxylase to fix carbon. Plants with the C_4 photosynthetic pathway have higher photosynthetic rates at current levels of atmospheric carbon dioxide; however, because additional energy is needed for the CO_2 pump, this type of photosynthesis gives plants the greatest advantage under conditions of high light and warmer temperatures. Because of the greater photosynthetic efficiency of the C_4 pathway, rising levels of atmospheric CO_2 are thought to have little direct effect on stimulating photosynthesis for C_4 plant species (Figure 3.2).

How would temperature alter the impact of rising CO_2 on the photosynthetic process? Temperature reduces the ratio of CO_2 to O_2 in solution, increasing the specificity of Rubisco for oxygenase and enhancing photorespiration. Because photorespiratory carbon loss (as CO_2) increases with temperature, and since rising carbon dioxide concentration shifts Rubisco activity toward carboxylase, rising CO_2 should stimulate net C_3 photosynthesis even more as temperature increases (e.g., Sage, 1995). However, long-term adaptation to growth temperature may lower the temperature sensitivity of CO_2-induced stimulations in C_3 photosynthesis (Bunce, 2000a; Ziska, 2001). For C_4 species, increasing temperature will likely stimulate photosynthesis (up to a species-dependent optimal temperature), but a synergistic effect between temperature and CO_2 is not anticipated because of the concentrating mechanism of C_4 photosynthesis.

Plants oxidize water as a source of electrons for photosynthesis, so water availability is a crucial aspect of carbon assimilation. Water stress is an acknowledged limitation on photosynthesis for both the C_3 and C_4 pathways. Stress can affect photosynthesis by reducing the stomatal aperture and limiting CO_2 availability, or through nonstomatal limitations, where the photosynthetic apparatus is directly affected. How much water stress affects photosynthesis indirectly by altering stomatal aperture, and how much it directly affects photosynthesis by nonstomatal means is still a subject of much discussion (Lawlor, 2002).

There are two other leaf-level processes that differ between C_3 and C_4 plants that are also of functional importance. First, C_4 plants need less Rubisco because of their CO_2-concentrating mechanism and need less nitrogen in their leaves because Rubisco is a principal leaf protein. As a result, nitrogen use efficiency (NUE) is, in general, greater in C_4 relative to C_3 leaves. A second consideration is water use efficiency (WUE), the ratio of leaf water loss during transpiration to carbon gained via photosynthesis. Because of their greater photosynthetic efficiency, C_4 leaves tend to have a higher WUE.

As CO_2 increases, NUE will increase in C_3 leaves, in part because investment in Rubisco will be less, and stomatal openings will narrow, reducing transpiration and increasing WUE for both C_3 and C_4 species. Improved WUE and leaf water content with increasing CO_2 helps C_3 and C_4 species maintain leaf photosynthesis with either increased salinity or decreased water availability. Indeed, there has been some suggestion that the greatest impact of CO_2 on C_4 photosynthesis is an indirect effect due to improved WUE and maintenance of leaf photosynthesis under water-limiting conditions (Drake and Leadley, 1991; Leakey et al., 2006). However, Ghannoum et al. (2000) have also suggested that the greater leaf temperatures resulting from CO_2-induced stomatal closure in C_4 leaves may be one factor that contributes directly to their greater photosynthetic rates under higher CO_2.

Whole Plant Responses

Because of the distinction in photosynthetic pathways at the cellular level, C_4 plants are, in general, likely to have the photosynthetic "edge" over C_3 species at subambient CO_2 concentrations and when light and temperature are high. This is consistent with the evolutionary history of C_4 photosynthesis, and the fact that they are found principally among tropical grasses in warm, high light regions, which reflects, in turn, the relative distribution of C_3 and C_4 plants (Figure 3.3).

But it is also important to consider the probable impacts of rising CO_2, temperature/water at the whole plant level for both C_3 and C_4 plants. Given that the exchange of carbon and

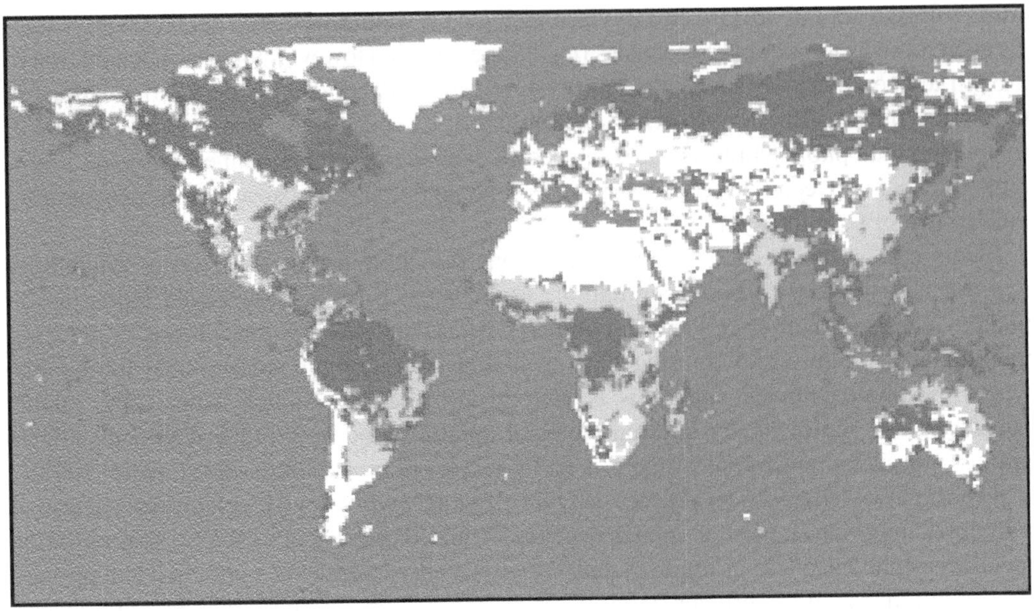

Figure 3.3 Distribution of plant biomes by photosynthetic pathway. C_4 grasslands (light grey) have evolved in the tropics and warm temperate regions where forests (dark grey) are excluded by seasonal drought or fire. C_3 plants (white) remain dominant in cool temperate grasslands because C_4 grasses are less productive at low temperatures.

water at the leaf level are critical to plant function, it is not surprising that growth of C_3 and C_4 plants increases with increasing photosynthesis and/or water efficiency (e.g., Patterson et al., 1999; Ziska and Bunce, 1997a). But the increased growth is not necessarily uniform, and allocation of carbon during growth is a critical feature of plant development and reproductive success. For a number of C_3 and C_4 plants, for example, increasing CO_2 may result in an increase in the ratio of root to shoot biomass (*R*:*S*), depending, in part, on nutrient stress and water availability (Rogers et al., 1992, 1994). As with increased root allocation, increased carbon allocation to reproduction has obvious implications with respect to seed production and viability.

An assessment of flowering time in relation to elevated atmospheric CO_2 showed that in the majority of cases, flowering times were accelerated; however, no differences were observed between C_3 and C_4 plants (Springer and Ward, 2007). Unfortunately, there is insufficient data to determine how CO_2-induced changes in flowering time could be influenced by other climatic factors such as temperature or water availability (Springer and Ward, 2007).

It has been suggested that plant development must proceed in such a way that growth is coordinated (e.g., Farrar and Gunn, 1996). In theory, partitioning of carbon to various sources or sinks will track carbon needs in changing environmental conditions. However, imbalances in source–sink relations in response to increasing carbon or abiotic stress can occur. Such imbalances can feed back to limit the ability of photosynthesis to respond to elevated carbon dioxide. For example, if air temperatures exceed the optimum for pollen formation, but not leaf function, then the resulting increase in floral sterility may limit reproductive sinks with a subsequent decline in the amount of photosynthetic stimulation in response to elevated CO_2 levels (e.g., Lin et al., 1997). If temperature and CO_2 enhance the rate of leaf development, then increased self-shading can occur also with a subsequent decline in whole plant photosynthesis as observed with increasing CO_2 and temperature for soybean (Ziska and Bunce, 1997b). The dynamics of source–sink relations and photosynthetic inhibitions are complex, but inhibition to rising CO_2 is not a ubiquitous response, even in the long term (Ainsworth et al., 2003). Overall, photosynthetic inhibition to elevated CO_2 is better characterized for C_3 plants; however, whether different feedback mechanisms are associated with a given photosynthetic type is unclear, and has not, to our knowledge, been studied.

Crop and Weed Biology: Revisiting a Paradigm

Many crops are descendant from weedy species and, as a result, are similar both genetically and ecologically. This is reflected in the overlap between congeneric weeds and crops. For example, it has been noted that several important weeds can also function as crop plants, depending on location (Holm et al., 1977). Overall, a number of common characteristics, including high growth rate, resource use, stress tolerance, and high reproductive output, can be found among herbaceous weeds and crops (Grime, 1979; Patterson, 1985).

We have focused on comparing the responses of C_3 and C_4 plants to rising CO_2 and climate because such a comparison has special relevance with respect to weeds and crops. One commonly accepted axiom in weed science is that among "worst" weeds, many possess the C_4 photosynthetic pathway, while most major crops are C_3 plants (Holm et al., 1977). Of the world's 18 worst weeds, 14 are C_4; conversely, of the dozen crops that constitute 75% of the world food supply, only 4 are C_4 plants (Holm et al., 1977).

Overall, it is tempting to view the differential responses among and between C_3 and C_4 plants as a means to predict the likely outcome between crops (C_3) and weeds (C_4) in the larger context of climate change. For example, an overall increase in atmospheric CO_2 would favor the C_3 crop over the C_4 weed; whereas increasing temperature, or reduced water availability could favor the C_4 weed. Much of the current scientific assessment in understanding the impact of climate change on weed and crop biology is, in fact, reflected in this C_3 versus C_4 paradigm (Bunce and Ziska, 2000; Patterson, 1986; Patterson et al., 1984; Patterson and Flint, 1980; Rosenzweig and Hillel, 1998).

However, it is worth examining, in closer detail, the source of this archetype. Dr. Leroy Holm was one of the early pioneers of weed science, and in a career spanning five decades, did exemplary work in cataloging the global distribution of many of the most pernicious weeds in agriculture for five continents. His three-volume set on "World's Worst Weeds," published in the 1970s, is familiar to almost all weed scientists.

It is this work, particularly in the context of agricultural commodities (e.g., wheat, corn, etc.), that distinguished Dr. Holm's research. His list of "worst weeds" included many C_4 species, including the number one worst weed, *Cyperus rotundus*, or purple nutsedge. Since

Table 3.1 The 15 most "serious" weeds as ranked and profiled in The World's Worst Weeds: Distribution and Biology (Holm et al., 1977) A = annual; P = perennial.

Rank	Common name	Scientific name	Form	C_3/C_4
1	Purple nutsedge	*Cyperus rotundus*	P	C_4
2	Bermudagrass	*Cynodon dactylon*	P	C_4
3	Barnyardgrass	*Echinochloa crus-galli*	A	C_4
4	Jungle rice	*Echinochloa colonum*	A	C_4
5	Goosegrass	*Eleusine indica*	A	C_4
6	Johnsongrass	*Sorghum halapense*	P	C_4
7	Cogongrass	*Imperata cylindrical*	P	C_4
8	Water hyacinth	*Eichhornia crassipes*	P	C_3
9	Common purslane	*Portulaca oleracea*	A	C_4
10	Lambsquarters	*Chenopodium album*	A	C_3
11	Large crabgrass	*Digitaria sanguinalis*	A/P	C_4
12	Field bindweed	*Convolvulus arvensis*	P	C_3
13	Wild oat	*Avena fatua*	A	C_3
14	Smooth pigweed	*Amaranthus hybridus*	A	C_4
15	Spiny amaranth	*Amaranthus spinosa*	A	C_4

many people are interested categorically in the idea of "worst" or "best," this particular species quickly gained the reputation as the world's worst weed (Table 3.1).

Yet, this information, invaluable as it was, is not based on any quantification of weed populations or objective criteria with respect to impact on crop production, but is described in Dr. Holm's words as:

> "... we have assembled and processed into meaningful form more data on the world weed problem than have been gathered up to this time, and *we must emphasize that the importance that we have assigned to each species is a personal judgment made against this background of experiment*. Our judgments must remain as fragile as yours until weed scientists in a future time can gather similar information and make other, more perfect judgments." [page xi, Preface to "Worlds Worst Weeds" (Italics are LHZ)]

Or, to summarize, the assessment of what constitutes "worst" is a personal judgment, reflecting both weed abundance and the capacity of the weed to inflict economic loss. Dr. Holm, to his credit, recognized that his opinion would change over time. Therefore, it is reasonable to ask whether, since the publication of "World's Worst Weeds," the C_4 photosynthetic pathway is still overrepresented among the worst, or most troublesome, weeds; as this has obvious implications with respect to rising carbon dioxide and climate change.

A limited comparison of "worst" for the United States reveals that the situation has indeed changed. What is interesting is that in comparing Holm's assessment with more recent assessments, there appears to be no overrepresentation of the C_4 pathway among "worst"

Table 3.2 Comparison of Holm's worst weeds and farmer assessment of worst weeds (Bridges, 1992) for corn, soybean, and wheat in the contiguous United States.

Crop	Holm et al. (1977)	Bridges (1992)
Corn	*Cyperus rotundus* (C_4)	*Sorghum halapense* (C_4), 18
	Digitaria sanguinalis (C_4)	*Ipomea* spp. (C_3), 14
	Echinochloa crus-galli (C_4)	*Albutilon theophrasti* (C_3), 14
	Sorghum halapense (C_4)	*Xanthium strumarium* (C_3), 11
	Portulaca oleracea (C_4)	*Amaranth* spp. (C_4), 11
	Cynodon dactylon (C_4)	*Chenopodium album* (C_3), 11
Soybean	*Eleusine indica* (C_4)	*Ipomea* spp. (C_3), 17
	Echinochloa colonum (C_4)	*Sorghum halapense* (C_4), 15
	Cyperus rotundus (C_4)	*Amaranths* spp. (C_4), 14
	Echinochloa crus-galli (C_4)	*Albutilon theophrasti* (C_3), 13
Wheat	*Avena fatua* (C_3)	*Cirsium arvense* (C_3), 13
	Polygonum convolvulus (C_3)	*Convolvulus arvensis* (C_3), 13
	Chenopodium alba (C_3)	*Bromus secalinus* (C_3), 11
	Convolvulus arvensis (C_3)	*Bromus tectorum* (C_3), 9

Listing, based on state noxious weed lists, of "worst weeds" for North America (28 states, plus 6 Canadian provinces, Skinner et al., 2000). Number indicates the number of states reporting the weed as the most troublesome.

1.	*Cirsium arvense* (C_3), 33 lists
2.	*Carduus nutans* (C_3), 24 lists
3.	*Lythrum* spp. (C_3), 24 lists
4.	*Convolvulus arvensis* (C_3), 23 lists
5.	*Euphorbia esula* (C_4), 22 lists
6.	*Acroptilon repens* (C_3), 20 lists
7.	*Sorghum* spp. (C_4), 20 lists
8.	*Cardaria* spp. (C_3), 17 lists
9.	*Centaurea maculosa* (C_3), 17 lists
10.	*Sonchus arvensis* (C_3), 17 lists
11.	*Centaurea diffusa* (C_3), 16 lists
12.	*Elytrigia repens* (C_3), 16 lists
13.	*Cuscuta* spp., 12 lists
14.	*Linaria dalmatica* (C_3), 12 lists
15.	*Centaurea solstitialis* (C_3), 11 lists

For Bridges, the number following the weed listing is the number of states reporting that weed as troublesome for the crop.

weeds in corn, soybeans, and wheat. In fact, for many growers in North America, C_3 weeds are now the "worst" weeds (13 out of 15; Skinner et al., 2000; Table 3.2).

We would emphasize then that any evaluation of what constitutes "worst," "worst" is subjective, depending on the perceptions of the farmer, grower, land manager, extension agent, etc. *at that moment*. No one review (e.g., Skinner et al., 2000) is, in fact, based on any sort of common objective criteria. This is due, in no small measure, to the time and cost needed to quantify the extent of crop losses due to a given weed species at the farm level.

Unfortunately, as of this writing, a comprehensive state-by-state evaluation of worst weeds for a given crop has not been conducted since the Bridges (1992) estimate.

It is also worth noting that although the most troublesome weeds in wheat are C_3, both Setaria and Amaranth spp. (C_4 weeds) are listed as troublesome weeds in wheat production in 6 and 8 states, respectively (Bridges, 1992). For both corn (C_4) and soybean (C_3), worst weeds include both C_3 and C_4 species. Consequently, the paradigm that C_3 plants are crops, C_4 plants are weeds, and that rising CO_2 will generally benefit crop production is an oversimplification of what is actually known regarding crop–weed associations. In fact, all major crops within the contiguous United States compete with an average of 8–10 "troublesome" weeds (Bridges, 1992), and no one crop competes solely with all C_3 or all C_4 weed species. In addition, such a paradigm does not address crop–weed associations where photosynthetic pathway is the same. Yet, in many instances, the worst or most troublesome weed is simply a wild relative of the crop (e.g., oat and wild oat, sorghum, and shattercane).

Evaluating Weed and Crop Responses: Rising Carbon Dioxide and the Environment

If the differential response of weeds and crops to rising atmospheric CO_2 and/or climate cannot be predicted by a simple photosynthetic categorization of weeds as C_4 and crops as C_3 species, what criteria can we use to determine the relative response of crops and weeds to rising atmospheric carbon dioxide? Are other biotic or abiotic factors likely to modify the degree of CO_2 stimulation?

David Patterson, one of the first scientists to investigate weed biology and climate change, classified the relative responses of a range of crops and weeds to a projected increase of ~300 ppm CO_2 above ambient for plants with the C_3 and C_4 type of photosynthesis. He found that C_3 and C_4 crops showed growth responses from 1.10 to 2.34× and 0.98 to 1.24×, whereas C_3 and C_4 weeds showed responses from 0.95 to 2.72× and 0.7 to 1.61×, compared to growth in ambient CO_2 concentrations (Patterson, 1993; Patterson and Flint, 1990; Patterson et al., 1999). This is, in many ways, a surprising result, given that at the biochemical level, C_4 plants should not respond at all to rising CO_2, let alone demonstrate such a broad range of responses. Similarly, why should C_3 species show such a broad range of enhancement to elevated CO_2? What is the basis for predicting the overall degree of photosynthesis or growth enhancement to rising levels of carbon dioxide among plants with the same photosynthetic pathway?

C_4 Weeds and Crops

Corn and sorghum are two principal C_4 crops, and based on early results, showed little photosynthetic or growth response to elevated CO_2 under well-watered conditions (e.g., Cure and Acock, 1986; Morrison and Gifford 1984; Patterson and Flint, 1980). These results were consistent with our basic understanding of the C_4 photosynthetic pathway and rising carbon dioxide (Berry and Downton, 1982).

However, with time, several studies reported increases in photosynthesis and growth of individual C_4 crops and weeds in response to elevated CO_2, even under well-watered conditions, examples included barnyardgrass and giant foxtail (Sionit and Patterson, 1984), redroot pigweed (Black, 1986; Ziska and Bunce, 1997a), itchgrass (Patterson and Flint, 1980), Bermudagrass (Ghannoum et al., 2001), and sorghum (Bunce, 2003; Cousins et al., 2001).

Although these studies indicate a direct effect of CO_2 on growth and/or photosynthesis in C_4 crops and weeds under well-watered conditions, the basis for any direct stimulation is unclear. The effect appears independent of different C_4 subtypes (Ziska and Bunce, 1997a), or the degree of CO_2 "leakiness" of the bundle sheath (Ziska et al., 1999) and is not related to any inhibition of dark respiration by rising CO_2 (Ziska and Bunce, 1999). There are some reports (Cousins et al., 2001; Moore et al., 1986) that indicate that a percentage of carbon is directly fixed by C_3 photosynthesis for leaves of C_4 species during early leaf development. This suggests that elevated CO_2 could potentially stimulate photosynthesis during leaf initiation, possibly due to the C_4 CO_2 concentrating apparatus being not yet fully developed. To date, however, no experiments have determined if CO_2 responsiveness is proportional to temporal changes in leaf development between C_4 monocots and C_4 dicots. Alternatively, there has been some suggestion that some C_4 weeds and crops may not be saturated at current CO_2 under field conditions, and could show a slight (10–15%) increase in photosynthesis and growth with increasing atmospheric carbon dioxide (Bunce, 2005; Ghannoum et al., 2000; Figure 3.4).

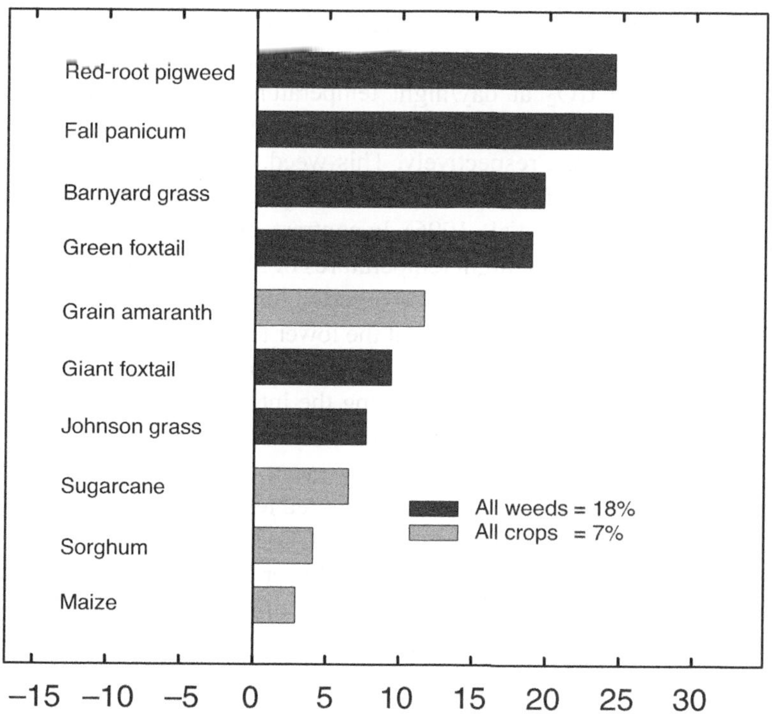

Figure 3.4 Relative response of C_4 weed and crop dry matter at projected levels of carbon dioxide (~690 vs. 385 ppm) at 60 days following emergence. (From Ziska and Bunce, 1997a.)

In contrast to any direct effect of CO_2 on photosynthesis, water shortages could also enhance the photosynthetic and growth response of C_4 weeds and crops to elevated CO_2. Plants reduce their stomatal aperture under elevated CO_2, losing less water and increasing transpiration efficiency. This could be important for the survival of C_4 plants in grassland environments, which often experience periods of limited water availability (e.g., Owensby et al., 1996); or, alternatively, if climate change lengthens periods between precipitation events (Wolfe et al., 2008). Several studies have indicated that this indirect effect of elevated CO_2 could stimulate photosynthesis and growth in C_4 weeds and crops during drought (e.g., Patterson, 1986 for barnyardgrass, goosegrass, and crabgrass; Leakey et al., 2004 for corn). However, there are also studies where elevated CO_2 did not enhance the water status of C_4 weeds under drought (e.g., broomsedge; Wray and Strain, 1987). Under many conditions, the indirect effect of CO_2 on stomatal closure could even enhance photosynthesis under well-watered conditions, since this would result in higher leaf temperatures (Ghannoum et al., 2000).

With respect to the potential interaction between CO_2 and temperature, if younger leaves did not have the fully developed C_4 pathway and could respond directly to rising CO_2, then rising temperatures could actually reduce photosynthetic stimulation by accelerating leaf development. However, the few data available on CO_2 and supraoptimal temperatures for C_4 weeds do not show any consistent interaction between these variables, with the effects being dependent on both species and populations. For example, Potvin and Strain (1985) examined three populations of the C_4 weed barnyardgrass collected over a latitudinal gradient from Mississippi to Quebec and found that the Quebec population demonstrated equal dry matter stimulation with elevated CO_2 at day/night temperatures of 28/22 and 21/15°C, while populations from North Carolina and Mississippi had greater growth responses to CO_2 at the higher and lower temperatures, respectively. This weed, which is among the *Echinochloa* spp. that are common weeds in flooded rice, did not respond to elevated CO_2 at two different day/night temperatures (Alberto et al., 1996). In contrast, photosynthesis of redroot pigweed responded less to CO_2 at high day/night temperatures of 38/31°C than at 28/22°C (Coleman and Bazzaz, 1992). Ackerly et al. (1992) also reported for this same experiment, an increase in leaf initiation and branch leaf production at the lower temperature with elevated CO_2, but not at the higher temperature, consistent with the photosynthetic results.

At present, almost nothing is known regarding the interaction of elevated CO_2 and sub-ambient temperatures in either C_4 weeds or crops. Potvin (1985) examined the response of barnyardgrass and goosegrass to a single chilling event (one night at 7°C) and found that plants grown at the higher CO_2 concentration exhibited less reduction in leaf photosynthetic rate and less visual chlorosis than plants at ambient CO_2.

Because C_4 species need less Rubisco, they have lower leaf protein concentrations than C_3 species, and higher nitrogen use efficiencies (amount of N per CO_2 molecule assimilated). Consequently, the sensitivity of C_4 crops and weeds to nutrient availability should be minimal. Redroot pigweed, for example, showed no interactive response between elevated CO_2 and nitrogen (Coleman et al., 1991). Similar patterns of response have also been reported for corn (Wong, 1979). However, for Japanese millet, enhancement of leaf area production and photosynthesis by elevated CO_2 occurred only at suboptimal NO_3 concentrations (Wong and Osmond, 1991). Other work has also suggested that lack of nitrogen could reduce the efficiency of the C_4 concentrating mechanism (Meinzer and Zhu, 1998), leading to a greater relative effect of rising CO_2 on C_4 species. However, characterization of the response of C_4 crops and weeds to rising CO_2 in the context of sub- or supraoptimal amounts of N is, at present, based on a small number of experiments, and generalizations are tentative. Similarly, little is known with respect to how other macronutrients (P or K), or other micronutrients (Fe and Mg) interact to modify CO_2 responses.

C_3 Weeds and Crops

As stated earlier in the chapter, the overwhelming majority (+95%) of plant species use the C_3 pathway. Although scientists have known for 200 years that enhanced levels of carbon dioxide stimulate the growth of these plants (De Saussure, 1804), the number of studies involving CO_2 exposure of agricultural and horticultural plants (almost all C_3 species) increased markedly in the 1960s and 1970s (e.g., Allen, 1979), following the discovery that global atmospheric CO_2 could be increasing (Keeling, 1960, see Chapter 2). By the early 1980s, Kimball (1983) had documented over 400 observations with respect to agricultural yield in response to elevated CO_2 conditions, data which could potentially be used to forecast the impact of rising carbon dioxide on crop productivity. In contrast, the first publications assessing weed responses to CO_2 were published by Patterson and Flint in 1980. In addition, the early focus was on C_3 species, and as a result there is a great deal of more information on the individual response of C_3 crops and weeds than for their C_4 counterparts.

Given what is known regarding C_3 photosynthesis, there is no question that C_3 weeds and crops will respond to rising CO_2 under optimal conditions. Experimental data has accumulated rapidly since the initial assessment by Kimball (1983) and with the advent of software to incorporate disparate results in meta-analyses, review papers accessing the response of C_3 crops are becoming more frequent (e.g., Ainsworth, 2008). These studies have found wide variation in the photosynthetic and growth responses of crops and weeds to rising CO_2. The basis for this variation is not entirely known. It has been suggested, for example, that the response of crops to CO_2 has been overestimated due to methodological artifacts (Long et al., 2006), or because agronomic conditions may be nonoptimal (Poorter and Perez-Soba, 2001), or because of ozone interactions (Heagle et al., 1999). However, any potential increase in carbon assimilation due to rising CO_2 and subsequent effects on growth and reproduction will depend on a number of abiotic factors that are likely to change with anthropogenic warming, including temperature, moisture, and nutrient availability. How these factors alter the individual responses of C_3 species to rising CO_2 is then an important consideration in elucidating differential morphological, phenological, allometric, or reproductive responses for crops and weeds at the whole plant level.

For a number of C_3 crop species, responses to CO_2 will depend on nutrient levels. For rice and wheat, any stimulation of crop yield by CO_2 will depend on N availability (e.g., Imai and Murata, 1978; Kim et al., 2003 for rice; Wolf, 1996 for wheat), whereas stimulation of yield by CO_2 in soybean appears independent of supplemental N (Cure et al., 1988). This may be related, in part, to sink development. Cereals need nitrogen for tiller formation early in development. Later, during maturity, these tillers can support a greater number of panicles, with seeds acting as a suitable sink for additional photosynthetic carbon and a greater amount of filled grain. By reducing these potential sinks, nitrogen-poor soils limit the photosynthetic response of cereals to CO_2. Less is known regarding the role of nutrients in the response of C_3 weeds to rising CO_2; although studies have shown that sicklepod, showy crotalaria, Canada thistle, and wild radish do not require additional nitrogen to respond to elevated CO_2 (Jablonski, 1997; Patterson and Flint, 1982; Ziska, 2003). Although the issue of nutrient availability in the context of plant growth with climate change has been examined at a number of levels (e.g., Reich et al., 2006; Sinclair, 1992), systematic comparisons between C_3 crops and weeds are rare. The available data suggest that CO_2 can stimulate crop and weed growth over a wide range of nutrient levels, although, in general, the greater growth enhancement with additional CO_2 is more likely to occur under optimal nutrient availability particularly for crops. The variability in the interaction between CO_2 responsiveness and nutrients for C_3 species has resulted in a reexamination of the idea of the "Law of the Minimum," which suggests plant growth is controlled by the resource in most limited supply. However, Sinclair (1992), inter alia, has suggested that nutrient resources can be simultaneously limiting, so that increasing one may compensate for limits imposed by others (Figure 3.5).

As discussed previously, at the biochemical level, the photosynthetic and growth responses of C_3 species to elevated CO_2 should be temperature-dependent. That is, the response to concurrent increases in CO_2 and temperature should exceed the response to CO_2 alone. Several reports indicate such a pattern for vegetative growth of crop species (Patterson, 1995), but the pattern for the reproductive response is less clear. For rice, increasing CO_2 and temperature may in fact, negate any yield enhancement related to increasing CO_2 (Moya et al., 1998; Ziska et al., 1996), due in part to reductions in transpirational cooling, higher canopy temperatures, and increased pollen sterility. The interaction between temperature and carbon dioxide seems more complex for wheat, with a majority of experiments indicating a reduction in yield with elevated CO_2 in combination with warming compared to elevated CO_2 alone (Amthor, 2001). A few studies have compared responses of soybean crops grown for the full season under different temperature regimes (Baker et al., 1989; Boote et al., 2005; Heinemann et al., 2006; Sionit et al., 1987) but have found only minor effects of these treatments on the yield response to elevated CO_2. These data indicate that temperature thresholds, particularly for flowering, may be significantly lower at elevated CO_2 in C_3 crops. If rising temperatures limit seed production, fewer sinks and a diminished photosynthetic response to CO_2 could occur as a result of feedback inhibition (e.g., Lin et al., 1997).

Conversely, fewer studies have examined the interactions of temperature and CO_2 for growth and reproductive responses in C_3 weeds. For the data that are available, some weed species show a greater growth enhancement to elevated CO_2 with warmer temperatures (spurred anoda and sicklepod), but other C_3 weeds do not (velvet leaf, quackgrass; Patterson et al., 1988; Tremmel and Patterson, 1993; Table 3.3).

In contrast to rising temperatures, little is known regarding the differential enhancement of C_3 crops and weeds to rising CO_2 at suboptimal temperatures. For okra, elevated CO_2 increased tolerance to low temperatures (Sionit et al., 1981); for dandelion, there were

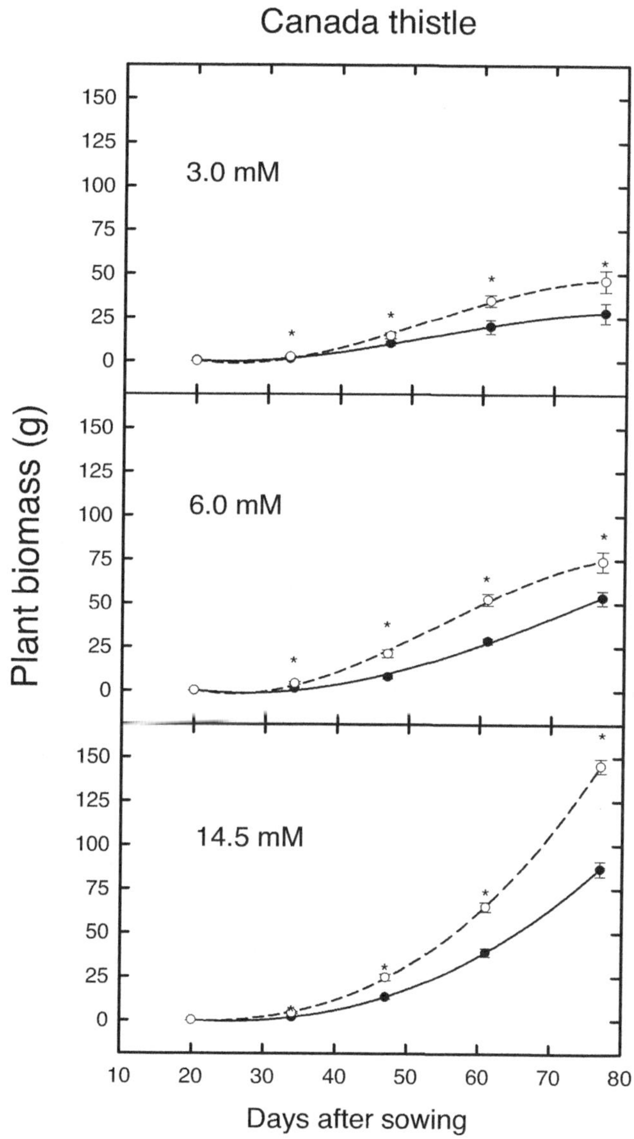

Figure 3.5 Change in plant biomass of Canada thistle (*Cirsium arvense*) from 20 through 77 days after sowing when grown at three levels of nitrogen and at 290 and 380 ppm CO_2 (beginning and end of the twentieth century, respectively). Although N supply affected growth rate, it did not change the relative effect of additional CO_2 on plant biomass. *indicates significant difference ($P < 0.05$, unpaired t-test) for a given sampling date. (From Ziska, 2003, *Physiologia Plantarum* 119:105–112.)

Table 3.3 Optimal air temperature ranges for vegetative and reproductive development, and failure temperature (floral temperature where yield falls to zero) for selected C_3 and C_4 crops.

Crop	C_3/C_4	Optimal temperature vegetative (°C)	Optimal temperature reproductive (°C)	Failure (°C)
Maize	C_4	NA	18–22	35
Soybean	C_3	25–37	22–24	39
Wheat	C_3	20–30	15–21	34
Rice	C_3	25–36	23–29	36
Sorghum	C_4	26–34	25	35
Cotton	C_3	30–35	25–26	35
Peanut	C_3	31–25	20–26	39

Data are taken from Table 2.3, USDA Synthesis and Assessment Product 4.3, except for rice which is taken from Ziska unpublished data. Similar data on C_3 and C_4 weeds are lacking.

no significant differences in the relative stimulation of photosynthesis by elevated carbon dioxide among days with temperatures ranging from 15 to 34°C (Bunce, 2000b). Bunce (2000b) suggested that it was premature to conclude that low temperature per se would limit crop or weedy responses to rising carbon dioxide.

Climate change is likely to increase precipitation in the northeastern United States, and decrease precipitation in the West and Southeast (IPCC, 2007). As discussed previously, elevated CO_2 can indirectly stimulate plant photosynthesis and growth by causing plants to narrow their stomatal openings, increasing water use efficiency. Reductions in evapotranspiration of a few percent could be important to C_3 crops and potential yields. Interestingly, although the interaction of CO_2 and water supply has been studied extensively for crop species (e.g., Bunce, 2004; Nijs et al., 1989; Tyree and Alexander, 1993), few studies have quantified differences between C_3 weeds and crops. In addition, little is known about responses of C_3 crops versus weeds to elevated CO_2 under flooding conditions, even though such conditions are anticipated for many regions with climate change (Wolfe et al., 2008).

Summary

Plant biological function varies temporally and spatially from the biochemical to the ecological. We have focused, for convenience's sake, on how the rise in carbon dioxide concentration is likely to differentially influence weeds and crops up to the level of the whole plant. We addressed the issue of "worst weeds," focusing on the suggestion by Leroy Holm and others that the C_4 photosynthetic pathway is overrepresented among the most troublesome weeds. Given the differential response between C_4 and C_3 weeds and crops to CO_2, and the research framework for labeling C_4 species as weeds, and C_3 species as crops, we have stressed, based on available data, how rising levels of CO_2 are likely to affect each photosynthetic type and how these impacts may be modified by concurrent changes in other climatic parameters. However, we also wish to emphasize that this paradigm is incomplete, and that, at the field level, C_3 crops and weeds compete with C_4 crops and weeds in ways that make predicting responses to rising CO_2 and/or climate difficult. In the next chapter, we will continue our analysis, but we will move beyond a simple assessment of photosynthetic pathway to examine how CO_2 and climate may alter biological processes that take place at higher functional levels such as competition and evolution.

References

Ackerly DD, Coleman JS, Morse SR, Bazzaz FA. 1992. CO_2 and temperature effects on leaf area production in two annual plant species. *Ecology* **73**:1260–1269.

Ainsworth EA. 2008. Rice production in a changing climate: a meta-analysis of responses to elevated carbon dioxide and elevated ozone concentration. *Global Change Biol.* **14**:1642–1650.

Ainsworth EA, Davey PA, Hymus GJ, Osborne CP, Rogers A, Blum H, Nosberger J, Long SP. 2003. Is stimulation of leaf photosynthesis by elevated carbon dioxide concentration maintained in the long term? A test with *Lolium perenne* grown for 10 years at two nitrogen fertilization levels under free air CO_2 enrichment (FACE). *Plant Cell Environ.* **26**:705–714.

Alberto AMP, Ziska LH, Cervancia CR, Manalo PR. 1996. The influence of increasing carbon dioxide and temperature on competitive interactions between a C_3 crop, rice (*Oryza sativa*) and a C_4 weed (*Echinochloa glabrescens*). *Aust. J. Plant Physiol.* **23**:795–802.

Allen LH Jr. 1979. Potential for carbon dioxide enrichment. In: Barfield BJ, Gerber JF (eds) *Modification of the Aerial Environment of Crops*. American Society Agricultural Engineers, St. Joseph, USA, pp. 500–519.

Amthor JS. 2001. Effects of atmospheric CO_2 concentration on wheat yield: review of results from experiments using various approaches to control CO_2 concentration. *Field Crops Res.* **73**:1–34.

Bae H, Sicher R. 2004. Changes of soluble protein expression and leaf metabolite levels in *Arabidopsis thaliana* grown in elevated atmospheric carbon dioxide. *Field Crops Res.* **90**:61–73.

Baker JT, Allen LH Jr., Boote KJ, Jones P, Jones JW. 1989. Response of soybean to air temperature and carbon dioxide concentration. *Crop Sci.* **29**:98–105.

Berry JA, Downton JS. 1982. Environmental regulation of photosynthesis. In: Govindgee (ed.) *Photosynthesis, Development, Carbon Metabolism and Plant Productivity*. Academic Press, New York, pp. 263–343.

Black CC. 1986. Effects of CO_2 concentration on photosynthesis and respiration of C_4 and CAM plants. In: Enoch HZ, Kimball BA (eds) *Physiology, Yield and Economics*. CRC Press, Boca Raton, FL, pp. 29–40.

Boote KJ, Allen LH Jr., Prasad PV, Baker JT, Gesch RW, Synder AM, Pan D, Thomas JM. 2005. Elevated temperature and CO_2 impact pollination, reproductive growth and yield of several globally important crops. *J. Agric. Met. Jap.* **60**:469–474.

Bridges DC. 1992. *Crop Losses Due to Weeds in the United States*. Weed Science Society of America, Champaign, USA, 403 pp.

Bunce JA. 2000a. Acclimation of photosynthesis to temperature in eight cool and warm climate herbaceous C_3 species: temperature dependence of parameters of a biochemical photosynthesis model. *Photosynth. Res.* **63**:59–67.

Bunce JA. 2000b. Acclimation to temperature of the response of photosynthesis to increased carbon dioxide concentration in *Taraxacum officinale*. *Photosynth. Res.* **64**:89–94.

Bunce JA. 2003. Effects of water vapor pressure difference on leaf gas exchange in potato and sorghum at ambient and elevated carbon dioxide under field conditions. *Field Crops Res.* **82**:37–47.

Bunce JA. 2004. Carbon dioxide effects on stomatal responses to the environment and water use by crops under field conditions. *Oecologia* **140**:1–10.

Bunce JA. 2005. What is the usual internal carbon dioxide concentration in C_4 species under midday field conditions? *Photosynthetica* **43**:603–608.

Bunce JA, Ziska LH. 2000. Crop ecosystem responses to climatic change: crop/weed interactions. In: Reddy KR, Hodges HF (eds) *Climate Change and Global Crop Productivity*. CABI Publishing, New York, pp. 333–348.

Cheng S-H, Moore BD, Seemann JR. 1998. Effects of short-and long-term elevated CO_2 on the expression of Ribulose-1,5-bisphosphate carboxylase/oxygenase genes and carbohydrate accumulation in leaves of *Arabidopsis thaliana* (L.) Heynh. *Plant Physiol.* **116**:715–723.

Coleman JS, Bazzaz FA. 1992. Effects of CO_2 and temperature on growth and resource use on co-occurring C_3 and C_4 annuals. *Ecology* **73**:1244–1259.

Coleman JS, Rochefort L, Bazzaz FA, Woodward FI. 1991. Atmospheric CO_2, plant nitrogen status and the susceptibility of plants to acute heat stress. *Plant Cell Environ.* **14**:667–674.

Cousins AB, Adam NR, Wall GW, Kimball BA, Pinter PJ Jr., Leavitt SW, LaMorte RL, Matthias AD, Ottman MH, Thompson TL, Webber AN. 2001. Reduced photorespiration and increased energy-use efficiency in young CO_2-enriched sorghum leaves. *New Phytol.* **150**:275–284.

Cure JD, Acock B. 1986. Crop responses to carbon dioxide doubling: a literature survey. *Agric. For. Meteorol.* **38**:127–145.

Cure JD, Israel DW, Rufty TW Jr. 1988. Nitrogen stress effects on growth and seed yield of non-nodulated soybean exposed to elevated carbon dioxide. *Crop Sci.* **28**:671–677.

De Saussure NT. 1804. *Recherches chimiques sur la vegetation* ("Chemical Researches on Vegetation"). Nyon Ed., Gauthiers-Villars, Paris, 1957 (reprinted from 1804).

Drake BG, Leadley PW. 1991. Canopy photosynthesis of C3 and C4 plant communities exposed lo long-term elevated CO2 treatment. *Plant Cell Environ.* **14**:853–860.

Farrar JF, Gunn S. 1996. Effects of temperature and atmospheric carbon dioxide on source/sink relations in the context of climate change. In: Zamski E, Shaffer A (eds) *Distribution of Photoassimilates in Plants and Crops*. Marcel Dekker, New York, pp. 389–406.

Ghannoum O, von Caemmerer S, Conroy JP. 2001. Plant water use efficiency of 17 Australian NAD-ME and NADP-ME C_4 grasses at ambient and elevated CO_2 partial pressure. *Aust. J. Plant Physiol.* **28**:1207–1210.

Ghannoum O, von Caemmerer S, Ziska LH, Conroy JP. 2000. The growth response of C_4 plants to rising atmospheric CO_2 partial pressure: a reassessment. *Plant Cell Environ.* **23**:931–942.

Grime JP. 1979. *Plant Strategies and Vegetation Processes*. John Wiley and Sons, New York, 222 pp.

Heagle AS, Booker FL, Miller JE, Pursley WA, Stefanski LA. 1999. Influence of daily CO_2 exposure duration and root environment on soybean responses to elevated CO_2. *J. Environ. Qual.* **28**: 666–675.

Heinemann AB, Maia A, Dourado-Neto D, Ingram KT, Hoogenboon G. 2006. Soybean (*Glycine max* Merr.) growth and development response to CO_2 enrichment under different temperature regimes. *Eur. J. Agron.* **24**:52–61.

Holm LG, Plucknett DL, Pancho JV, Herberger JP. 1977. *The Worlds Worst Weeds*. University of Hawaii Press, Honolulu, HI, 609 pp.

Imai K, Murata Y. 1978. Effects of carbon dioxide concentration on growth and dry matter production of crop plants. III. Relationship between CO_2 concentration and nitrogen nutrition in some C_3 and C_4 species. *Jpn. J. Crop. Sci.* **47**:118–123.

IPCC: Climate Change 2007. *Impacts, Adaptation and Vulnerability*. IPCC Secretariat, Geneva, Switzerland, 986 pp.

Jablonski LM. 1997. Responses of vegetative and reproductive traits to elevated CO_2 and nitrogen in *Raphanus* varieties. *Can. J. Bot.* **75**:533–545.

Keeling CD. 1960. The concentration and isotopic abundances of carbon dioxide in the atmosphere. *Tellus* **12**:200–203.

Kim HY, Lieffering M, Kobayashi K, Okada M, Miura S. 2003. Seasonal changes in the effects of elevated CO_2 on rice at three levels of nitrogen supply: a free air CO_2 enrichment (FACE) experiment. *Global Change Biol.* **9**:826–837.

Kimball BA. 1983. Carbon dioxide and agricultural yield: an assemblage and analysis of 430 prior observations. *Agron. J.* **75**:779–788.

Kimball BA, Mauney JR, Nakayama FS, Idso SB. 1993. Effects of increasing atmospheric CO_2 on vegetation. *Vegetation* **104/105**:65–75.

Koch KE. 1996. Carbohydrate modulated gene expression in plants. *Ann. Rev. Plant Physiol. Plant Mol. Biol.* **47**:509–540.

Lawlor DW. 2002. Limitation to photosynthesis in water stressed leaves: stomata vs. metabolism and the role of ATP. *Ann. Bot.* **89**:871–885.

Leakey ADB, Bernacchi CJ, Dohleman FG, Ort DR, Long SP. 2004. Will photosynthesis of maize (Zea mays) in the US corn belt increase in future CO_2 rich atmospheres? An analysis of diurnal courses of CO_2 uptake under free-air concentration enrichment (face). *Global Change Biol.* **10**:951–962.

Leakey ADB, Uribelarrea M, Ainsworth EA, Naidu SL, Rogers A, Ort DR, Long SP. 2006. Photosynthesis, productivity and yield of maize are not affected by open-air elevation of CO_2 concentration in the absence of drought. *Plant Physiol.* **140**:779–790.

Lin W, Ziska LH, Namuco OS, Bai K. 1997. The interaction of high temperature and elevated CO_2 on photosynthetic acclimation of single leaves of rice *in situ*. *Physiol. Plant* **99**:178–184.

Long SP, Ainsworth EA, Leakey ADB, Nosberger J, Ort DR. 2006. Food for thought: lower-than-expected crop yield stimulation with rising CO_2 concentrations. *Science* **312**:1918–1921.

Ludewig F, Sonnewald U. 2000. High CO_2-mediated down-regulation of photosynthetic gene transcripts is caused by accelerated leaf senescence rather than sugar accumulation. *FEBS Lett.* **479**:19–24.

Lüthi D, Floch ML, Bereiter B, Blunier T, Barnola J-M, Siegenthaler U, Raynaud D, Jouzel J, Fischer H, Kawamura K, Stocker TF. 2008. High-resolution carbon dioxide concentration record 650,000–800,000 years before present. *Nature* **453**:379–382.

Makino A, Nakano H, Mae T, Shimada T, Yamamoto N. 2000. Photosynthesis, plant growth and N allocation in transgenic rice plants with decreased rubisco under CO_2 enrichment. *J. Exp. Bot.* **50**:383–389.

Meinzer FC, Zhu J. 1998. Nitrogen stress reduces the efficiency of the C_4 CO_2 concentrating system, and therefore quantum yield, in Saccharum (sugarcane) species. *J. Exp. Bot.* **49**:1227–1234.

Moore BD, Cheng SH, Edwards GE. 1986. The influence of leaf development on the expression of C_4 metabolism in Flaveria trinervia, a C_4 dicot. *Plant Cell Physiol.* **27**:1159–1167.

Moore BD, Cheng S-H, Sims D, Seemann JR. 1999. The biochemical and molecular basis for photosynthetic acclimation to elevated atmospheric CO_2. *Plant Cell Environ.* **22**:567–582.

Morrison JIL, Gifford RM. 1984. Plant growth and water use with limited water supply in high CO_2 concentrations. II. Plant dry weight, partitioning and water use efficiency. *Aust. J. Plant Physiol.* **11**:375–384.

Moya TB, Ziska LH, Namuco OS, Olszyk DA. 1998. Growth dynamics and genotypic variation in tropical, field grown paddy rice (Oryza sativa L.) in response to increasing carbon dioxide and temperature. *Global Change Biol.* **4**:645–656.

Nijs I, Imprens I, Behaege T. 1989. Effects of different CO_2 environments on the photosynthesis-yield relationship and the carbon and water balance of a whole clover (*Trifolium repens* L. cv. Blanca) sward. *J. Exp. Bot.* **40**:353–359.

Owensby CE, Ham JM, Knapp A, Rice CW, Coyne PI, Auen LM. 1996. Ecosystem level responses of tallgrass prairie to elevated CO_2. In: Koch GW, Mooney HA (eds) *Carbon Dioxide and Terrestrial Ecosystems*. Academic Press, San Diego, CA, pp. 147–162.

Patterson DT. 1985. Comparative ecophysiology of weeds and crops. In: Duke SO (ed.) *Weed Physiology, Vol. 1, Reproduction and Ecophysiology*. CRC Press, Inc., Boca Raton, FL, pp. 101–129.

Patterson DT. 1986. Responses of soybean (*Glycine max*) and three C_4 grass weeds to CO_2 enrichment during drought. *Weed Sci.* **34**:203–210.

Patterson DT. 1993. Implications of global climate change for impact of weeds, insects and plant diseases. *Int. Crop Sci.* **1**:273–280.

Patterson DT. 1995. Effects of environmental stress on weed/crop interactions. *Weed Sci.* **43**:483–490.

Patterson DT, Flint EP. 1980. Potential effects of global atmospheric CO_2 enrichment on the growth and competitiveness of C_3 and C_4 weed and crop plants. *Weed Sci.* **28**:71–75.

Patterson DT, Flint EP. 1982. Interacting effects of CO_2 and nutrient concentration. *Weed Sci.* **30**:389–394.

Patterson DT, Flint EP. 1990. Implications of increasing carbon dioxide and climate change for plant communities and competition in natural and managed ecosystems. In: Kimball BA, Rosenburg NJ, Allen LH Jr. (eds) *Impact of Carbon Dioxide, Trace Gases and Climate Change on Global Agriculture*. ASA Special Publication No. 53. American Society of Agronomy, Madison, WI, pp. 83–110.

Patterson DT, Flint EP, Beyers JL. 1984. Effects of CO_2 enrichment on competition between a C_4 weed and a C_3 crop. *Weed Sci.* **32**:101–105.

Patterson DT, Highsmith MT, Flint EP. 1988. Effects of temperature and CO_2 concentration on the growth of cotton (*Gossypium hirsutum*), spurred anoda (*Anoda cristata*), and velvetleaf (*Abutilon theophrasti*). *Weed Sci.* **36**:751–757.

Patterson DT, Westbrook JK, Joyce RJV, Rogasik J. 1999. Weeds, insects and diseases. *Clim. Change* **43**:711–727.

Pearson PN, Palmer MR. 2000. Atmospheric carbon dioxide concentrations over the past 60 million years. *Nature* **406**:695–699.

Poorter H, Navas M-L. 2003. Plant growth and competition at elevated CO_2: on winners, losers and functional groups. *New Phytol.* **157**:175–198.

Poorter H, Perez-Soba M. 2001. The growth response of plants to elevated CO_2 under non-optimal conditions. *Oecologia* **129**:1–20.

Potvin C. 1985. Amelioration of chilling effects by CO_2 enrichment. *Physiologie Végétale* **23**:345–352.

Potvin C, Strain BR. 1985. Effects of CO_2 enrichment and temperature on growth in two C_4 weeds, *Echinochloa crus-galli, and Eleusine indica. Can. J. Bot.* **63**:1495–1499.

Reich PB, Hungate BA, Luo Y. 2006. Carbon-nitrogen interactions in terrestrial ecosystems in response to rising atmospheric carbon dioxide. *Annu. Rev. Ecol. Syst.* **37**:611–636.

Rogers HH, Peterson CM, McCrimmon JN, Cure JD. 1992. Response of plant roots to elevated atmospheric carbon dioxide. *Plant Cell Environ.* **15**:749–752.

Rogers HH, Runion GB, Krupa SV. 1994. Plant responses to CO_2 enrichment with emphasis on roots and the rhizosphere. *Environ. Pollut.* **83**:155–189.

Rosenzweig C, Hillel D. 1998. Effects on weeds, insects and diseases. In: Rosenzweig C, Hillel D (eds) *Climate Change and the Global Harvest*. Oxford University Press, Cambridge, UK, pp. 101–122.

Sage RF. 1995. Was low atmospheric CO_2 during the Pleistocene a limiting factor for the origin of agriculture? *Global Change Biol.* **1**:93–106.

Sheen J. 1994. Feedback control of gene expression. *Photo. Res.* **39**:427–438.

Siegenthaler U, Stocker TF, Monnin E, Lüthi D, Schwander J, Stauffer B, Raynaud D, Barnola J-M, Fischer H, Masson-Delmotte V, Jouzel J. 2005. Stable carbon cycle-climate relationship during the late Pleistocene. *Science* **310**:1313–1317.

Sinclair TR. 1992. Mineral nutrition and plant growth response to climate change. *J. Exp. Bot.* **43**:1141–1146.

Sionit N, Patterson DT. 1984. Responses of C_4 grasses to atmospheric CO_2 enrichment. I. Effect of irradiance. *Oecologia* **65**:30–34.

Sionit N, Strain BR, Beckford HA. 1981. Environmental controls on the growth and yield of okra. I. Effects of temperature and of CO_2 enrichment at cool temperature. *Crop Sci.* **21**:885–888.

Sionit N, Strain BR, Flint EP. 1987. Interaction of temperature and CO_2 enrichment on soybean: photosynthesis and seed yield. *J. Plant Sci.* **67**:629–636.

Skinner K, Smith L, Rice P. 2000. Using noxious weed lists to prioritize targets for developing weed management strategies. *Weed Sci.* **48**:640–644.

Solomon S, Qin D, Manning M, Alley RB, Berntsen T, Bindoff NL, Chen Z, Chidthaisong A, Gregory JM, Hegerl GC, Heimann M, Hewitson B, Hoskins BJ, Joos F, Jouzel J, Kattsov V, Lohmann U, Matsuno T, Molina M, Nicholls N, Overpeck J, Raga G, Ramaswamy V, Ren J, Rusticucci M, Somerville R, Stocker TF, Whetton P, Wood RA, Wratt D. 2007. Technical summary. In: Solomon S, Qin D,

Manning M, Chen Z, Marquis M, Averyt KB, Tignor M, Miller HL (eds) *Climate Change 2007: The Physical Science Basis. Contribution of Working Group I to the Fourth Assessment Report of the Intergovernmental Panel on Climate Change.* Cambridge University Press, Cambridge, UK, pp. 131–217.

Springer CJ, Ward JK. 2007. Flowering time and elevated atmospheric CO_2. *New Phytol.* **176**:243–255.

Tremmel DC, Patterson DT. 1993. Responses of soybean and five weeds to CO_2 enrichment under two temperature regimes. *Can. J. Plant Sci.* **73**:1249–1260.

Tyree MT, Alexander JD. 1993. Plant water relations and the effects of elevated CO_2: a review and suggestions for future research. *Vegetation* **104/105**:47–62.

Wolf J. 1996. Effects of nutrient supply (NPK) on spring wheat response to elevated atmospheric CO_2. *Plant Soil* **185**:113–123.

Wolfe DW, Ziska LH, Petzoldt C, Seaman A, Chase L, Hayhoe K. 2008. Projected change in climate thresholds in the Northeastern U.S.: implications for crops, pests, livestock and farmers. *Mitig. Adapt. Strategies Glob. Chang.* **13**:607–613.

Wong SC. 1979. Elevated atmospheric partial pressure of CO_2 and plant growth. I. Interactions of nitrogen and photosynthetic capacity in C3 and C4 plants. *Oecologia* **44**:68–74.

Wong SC, Osmond CB. 1991. Elevated atmospheric partial pressure of CO_2 and plant growth. III. Interaction between Triticum aestivum (C_3) and Echinochloa frumentacea (C_4) during growth in a mixed culture under different CO_2, N nutrition and irradiance treatments with emphasis on below ground responses estimated using the δ13C value of root biomass. *Aust. J. Plant Physiol.* **18:**137–152.

Wray SM, Strain BR. 1987. Response of two old field perennials to interactions of CO_2 enrichment and drought stress. *Am. J. Bot.* **73**:1486–1491.

Ziska LH. 2001. Growth temperature can alter the temperature dependent stimulation of photosynthesis by elevated carbon dioxide in *Albutilon theophrasti*. *Physiol. Plant* **111**:322–328.

Ziska LH. 2003. The impact of nitrogen supply on the potential response of a noxious, invasive weed, Canada thistle (Cirsium arvense) to recent increases in atmospheric carbon dioxide. *Physiol. Plant* **119**:105–112.

Ziska LH, Bunce JA. 1997a. Influence of increasing carbon dioxide concentration on the photosynthetic and growth stimulation of selected C_4 crops and weeds. *Photo. Res.* **54**:199–208.

Ziska LH, Bunce JA. 1997b. The role of temperature in determining the stimulation of CO_2 assimilation at elevated carbon dioxide in soybean seedlings. *Physiol. Plant* **100**:126–132.

Ziska LH, Bunce JA. 1999. Effect of elevated carbon dioxide concentration at night on the growth and gas exchange of selected C_4 species. *Aust. J. Plant Physiol.* **26**:71–77.

Ziska LH, Manalo PA, Ordonez RA. 1996. Intraspecific variation in the response of rice (*Oryza sativa* L.) to increased CO_2 and temperature: growth and yield response of 17 cultivars. *J. Exp. Bot.* **47**:1353–1359.

Ziska LH, Sicher RC, Bunce JA. 1999. The influence of elevated carbon dioxide on the growth and gas exchange of three C_4 species differing in decarboxylation type. *Physiol. Plant* **105**:74–80.

4 An Evaluation of the Impact of Rising Carbon Dioxide and Climatic Change on Weed Biology: Competition to Community Composition

"..... methinks, I would not grow so fast, because sweet flow'rs are slow and weeds make haste."

William Shakespeare

Introduction

In the previous chapter, we began to examine how (or whether) C_3 and C_4 weeds and crops could respond to carbon dioxide within agricultural systems, and why any growth and/or reproductive response may reflect not only the photosynthetic pathway, but also potential climatic interactions including temperature, nutrients, and water.

In this chapter, we will take a broader approach, evaluating biological processes beyond the level of the whole plant that are likely to be impacted by CO_2 and climate change, including competitive outcomes as well as potential changes in community composition and ecosystems. Our objective in doing so is to provide insight into how climate and CO_2 will alter inter- and intraspecific competition; potential demographic changes in plant community composition; and changes in ecosystem processes, including a possible role for climate and/or CO_2 in evolutionary selection among weed species.

Weeds and Competition

Defining Competition

A standard dictionary definition of competition would include the terms: "rivalry" and "opposition" in describing individuals that seek a common goal simultaneously. Such a definition does not entirely capture what is meant by plant competition, however. Plant competition is usually defined as occurring when the level of a physical resource(s) is less than that required to meet the physiological demands of two or more individuals. Competition among individuals of the same species and between individuals of different species is known as intra- and interspecific competition, respectively.

Competition and Early Establishment

One of the first competitive interactions follows seedling emergence, when light competition between and within species is a primary growth constraint. Indeed, measurements of leaf area following emergence can be used as a simple predictor of crop–weed competition (Kropff and Spitters, 1991). Changes in temperature and moisture projected with changing climates (IPCC, 2007) have obvious implications for germination and the spatial and temporal emergence of weed seeds. For example, dormancy, one of the major constraints on weed seed emergence, is likely to be broken by greater moisture and warmer temperatures; similarly, rates of seed herbivory or decomposition are also likely to be affected.

In agronomic situations, soil temperature and moisture are important in determining optimal planting times, which can give crops a "head start" in emergence relative to weeds. Temperature is the single most important factor regulating germination of nondormant seeds in irrigated, annual agroecosystems under conditions where light, nutrients, and moisture may be nonlimiting. The use of temperature and time in a degree-day approach has been used successfully in predicting phenological stages for a number of crop and weed species (Alm et al., 1991). A basic requirement for this approach is an estimate of the temperature below which phenological development ceases (T_{base}) for each species. The range of temperature baselines for phenological development among weed species can vary considerably. This means that weeds, collectively, are relatively insulated against climate variability, particularly with respect to milder winters and early springs. Crops, in contrast, are typically planted one variety at a time, and so germination and early development of crops are more likely to suffer from increased climatic variability (Table 4.1).

Atmospheric carbon dioxide concentrations may also influence seed germination and emergence rates. Increased CO_2 enhances seed germination of some herbaceous (Esashi et al., 1987; Heichel and Jaynes, 1974) and woody species (Mohan et al., 2004). It can be argued that, because soil CO_2 concentrations already far exceed those of the atmosphere, seeds would not perceive any changes in their environment. However, surface levels of

Table 4.1 Baseline temperature (°C, T_{base}) for phenological development for selected winter and summer annual weeds.

Weed			
Common name	Scientific name	Winter/Summer	T_{base}
Prostrate pigweed	*Amaranthus albus*	Summer	15.5
Palmer Amaranth	*Amaranthus palmeri*	Summer	17.1
Common crabgrass	*Digitaria sanguinalis*	Summer	15.1
Barnyard grass	*Echinochloa crus-galli*	Summer	13.0
Purslane	*Portulaca oleracea*	Summer	8.5
Yellow foxtail	*Setaria glauca*	Summer	11.7
Canadian horseweed	*Conyza canadensis*	Winter/Summer	12.5
Greek mustard	*Hirshfeldia incana*	Winter	10.0
Common sowthistle	*Sonchus oleraceus*	Winter	5.3

Adapted from Steinmaus et al. (2000)

CO_2 can be dynamic, ranging from ambient to several thousand parts per million, and the influence of increased atmospheric CO_2 on surface concentrations has not been well documented. Certainly, stimulation of seed germination with enhanced CO_2 does not always occur (e.g., Garbutt et al., 1990). However, increasing CO_2 could interact with or increase the production of ethylene, a plant growth regulator that stimulates seed germination (e.g., Esashi et al. 1987).

If warming temperatures, altered precipitation, or increased CO_2 alter emergence times, this will significantly affect competition. The first species to emerge, whether crop or weed, has a great competitive advantage. Firbank and Watkinson (1985) demonstrated that emergence time and density of seedling emergence could account for up to 50% of the variation in aboveground biomass of individual corn cockle plants. Ziska and Bunce (1993) found that warming and CO_2 increased germination and emergence of one crop (alfalfa) and three weedy species, suggesting potential shifts in postemergence and light competition between crops and weeds. Although these initial data are intriguing, it is not yet clear whether germination and emergence of weeds and crops respond differently to CO_2 and/or climate.

Competition and Growth

Plant growth rates following emergence also have a strong influence on competitive ability. For example, if a species grows tall quickly and shades other species it will have a competitive advantage over those plants below it. Similarly, rapid root growth and nutrient uptake may also confer a competitive advantage. Growth, in turn, will depend on environment, genetics, and phenotypic plasticity.

As a consequence, anthropogenic changes in the environment such as CO_2, temperature, precipitation, etc. are likely to alter growth rates differentially, due in part to species-specific effects at the biochemical or leaf level. Such differential growth responses, in turn, are almost certain to alter competitive outcomes at the inter- or intraspecific level (e.g., Vengris et al., 1955). A wide range of growth responses has been observed for annual and perennial weed species, both with respect to recent increases in CO_2 as well as those projected for the end of the current century (Table 4.2).

It should be emphasized, however, that it is difficult to predict responses among plant species sharing a given physiological or morphological trait to a change in an environmental variable. For example, because of the Holm et al. (1977) observation that "worst" weeds favored the C_4 pathway, many early experiments compared the growth of C_3 crop and C_4 weed species at elevated CO_2 even though the plants did not compete directly. Weed/crop biomass ratios were decreased at elevated CO_2 in itchgrass (C_4)/soybean (C_3) comparisons (Patterson and Flint, 1980; Patterson et al., 1979), and barnyardgrass, goosegrass, and southern crabgrass (all C_4 species)/soybean (C_3) comparisons (Patterson, 1986). Many additional examples could be gleaned from the published literature, but they are only suggestive of how C_3/C_4 comparisons might change with increased carbon dioxide. This is due, in part, to the dynamic of multispecific community response to CO_2 being difficult to predict based on the response of individual species to CO_2. This is emphasized by the work of Bazzaz and coworkers (Bazzaz and Carlson, 1984; Bazzaz and Garbutt, 1988; Zangerl and Bazzaz, 1984) who

Table 4.2 Overview of potential growth response of selected weed species to either recent or projected changes in the concentration of atmospheric carbon dioxide CO_2.

Species	Common name	Recent	Projected	Reference
Albutilon theophrasti	Velvetleaf	1.3–1.58	0.88–2.29	Bazzaz et al. (1989), Bunce (2001), Dippery et al. (1995), Morose and Bazzaz (1994)
Ambrosia artemisiifolia	Annual ragweed	NA	1.22–1.57	Bazzaz and Carlson (1984), Hirose et al. (1996)
Avena fatua	Wild oat	NA	1.84	Freeden and Field (1995)
Brassica kaber	Wild mustard	NA	1.27–1.30	Wayne et al. (1999)
Bromus madritensis	Red brome	NA	1.09–1.51	Huxman et al. (1999.), Smith et al. (2000)
Bromus tectorum	Cheatgrass	NA	1.39–1.93	Poorter (1993), Smith et al. (1987)
Centaurea maculosa	Spotted knapweed	2.15	1.46	Ziska (2003a)
Centaurea solstitialis	Yellow starthistle	1.87	1.15–1.68	Dukes (2002), Poorter et al. (1996), Ziska (2003a)
Cirsium arvense	Canada thistle	1.58–2.84	1.69–1.70	Ziska (2002, 2003a, 2003b)
Convolvulus arvensis	Field bindweed	1.85	1.36	Ziska (2003a)
Datura stromonium	Jimsonweed	NA	1.76	Garbutt and Bazzaz (1984)
Elytrigia repens	Quackgrass	NA	1.12–1.68	Tremmel and Patterson (1993), Ziska and Teasdale (2000)
Euphorbia esula	Leafy spurge	1.95	1.43	Ziska (2003a)
Hydrilla verticillata	Waterthyme	NA	1.04–1.46	Chen et al. (1994.)
Lonicera japonica	Japanese honeysuckle	NA	2.34–3.60	Belote et al. (2003), Sasek and Strain (1988)
Prosopis glandulosa	Honey mesquite	1.53	1.11–1.37	Polley et al. (1994, 1996, 2002)
Pueraria montana	Kudzu	NA	1.20–2.15	Sasek and Strain (1991)
Sonchus arvensis	Perennial sowthistle	2.09	1.66	Ziska (2003a)

Recent changes refer to the increase in CO_2 from 250–300 ppm (sub-ambient) to 360–400 ppm (current ambient concentration); projected changes refer to the increase in CO_2 from 360–400 ppm to 600–800 ppm, a value expected to occur before the end of the current century. Values given are the ratio of plant biomass produced at the two CO_2 comparisons (i.e., a value of 1 indicates no response to increasing CO_2 while a value of 2 indicates a doubling in biomass). NA, not available. Data are from Ziska and George (2004).

measured species response to CO_2 when plants were grown alone or in competition and found little relationship between the relative growth rates of isolated and competing plants. For example, it was found that the red-root pigweed, a C_4 weed, out competed velvetleaf, a C_3 weed, in response to rising carbon dioxide (Bazzaz et al., 1989). Similarly, it cannot be assumed that rising temperature will uniformly favor C_4 species in competitive situations. Although there is evidence that C_4 species gain a photosynthetic advantage in warm, high light environments, C_4 species such as pigweed, barnyardgrass, cordgrass, and common purslane also thrive in the cooler climates of Canada and northern Europe.

Allocation of the additional carbon made available with rising CO_2 will have consequences for competition. For example, various vine species, including a number of well-known weed species (e.g., kudzu, Japanese honeysuckle) grow and photosynthesize much faster under CO_2 enrichment (Condon et al., 1992; Sasek and Strain, 1990, 1991). This dramatic response may reflect a positive feedback of high CO_2 for vines; that is, with an increase in CO_2 and a corresponding increase in photosynthesis, vines can allocate more photosynthate to additional leaves (i.e., more photosynthetic tissue), because they need less support tissue than other woody species (Putz and Mooney, 1991; Sasek and Strain, 1989; Schnitzer and Bongers, 2002). As a result, there is initial evidence that an increasing abundance of woody vines is causing increased tree mortality and reduced tree regeneration in forests around the globe (Dillenburg et al., 1995; Laurance et al., 2001; Phillips et al., 2002). It can be argued that such differential responses of growth to rising CO_2 are nutrient- or water-dependent (e.g., Reich et al., 2006). However, in many agronomic situations involving crops and weeds, nutrient and water availability may be optimal, potentially promoting the growth of vine-like weeds such as field bindweed.

Differential growth responses to climate change may also alter competitive outcomes indirectly by changing rates of herbivory. For example, Canada thistle responds to rising CO_2 by increasing the number or size of leaf spines (Ziska et al., 2004a). For populations of cheatgrass, a ubiquitous weed of western rangelands, the indigestible portion of aboveground plant material (acid detergent fiber [ADF], mostly cellulose and lignin) increased with increasing CO_2 at 50 ppm intervals from 270 to 420 ppm (Ziska et al., 2005). Differential effects of climate- and/or CO_2-induced changes in physical defenses or palatability have consequences for growth, depending on species. Recent data suggest that growth at elevated carbon dioxide compromised the ability of soybean plants to defend against insect predation (Zavala et al., 2008). It is not clear as how many crops and weeds will respond in this manner. Similarly, although it has been shown that some weed species may gain a competitive advantage via allelopathy (e.g., catechin in spotted knapweed; Thelen et al., 2005), the impact of climate on allelopathy, and possible differential effects on plant-to-plant competition are unknown.

Competition and Reproduction

Ultimately, "winning" and "losing" plant species are determined by the number of progeny added to the environment, and the subsequent persistence and presence of those individuals in the plant community over time. To that end, germination, growth, and efficient capture of physical resources are all necessary in order to maximize reproduction. How then is climatic change likely to alter sexual and asexual production within weed species?

Sexual reproduction occurs with transfer of male DNA to the female flowers. Male DNA in pollen can be transported via wind (anemophily), by insects (entomophily), or by vertebrates (zoophily). Many weed species are anemophilous, including most grasses, pigweed, lambsquarters, and ragweed. Although wind can transport pollen, it is not an efficient vector, and large quantities of pollen must be produced. For common ragweed, rising carbon dioxide has been shown to increase pollen production (Wayne et al., 2002; Ziska and Caulfield, 2000) and pollen quality (Singer et al., 2005) and warmer temperatures ($\sim$1.2°C above ambient) may also increase pollen production and size from western ragweed (Wan et al.,

2002). Urban-induced changes in CO_2 concentration, growing season length, and warmer temperatures were also associated with greater pollen production, as well as earlier flowering (Ziska et al., 2003). In addition, there are clear effects of CO_2 and temperature on plant height for common ragweed and lambsquarters with potential effects on the range of pollen dispersal (Ziska et al., 2004b). Similar changes in regard to pollen production and dispersal are certainly possible for other anemophilous weeds, but have not been well studied. Given that fewer crop species are wind-pollinated, such changes could promote the presence of anemophilous weeds (relative to crops) in the environment.

Few entomophilous weeds are host-specific, instead relying on a number of insect vectors rather than a single insect pollinator. Weeds attract generalist pollinators by producing flowers that are nectar rich. Examples include milkweed, Japanese honeysuckle, and goldenrod. In contrast to weeds, many crop species, including commercial plantings of cucumbers, squash, melons, strawberries, etc. rely on European honeybees as the principal pollinator.

How then is climate change likely to impact insect vectors? In regard to climate change, there is a clear signal for warmer winters and earlier springs, with advancement of flowering times (IPCC, 2007). Several studies have examined whether anthropogenic climate change can alter the phenological development of specific insect vectors and whether their mutualisms will become spatially or temporally misaligned in the future (Fitter and Fitter, 2002; Gordo and Sanz, 2005; Parmesan et al., 1999). Empirically, given this type of potential disruption, those plant species that rely on generalist pollinators are likely to be at a competitive advantage. In addition, crops may be particularly sensitive to temperature impacts on pollen sterility (Hatfield et al., 2008). However, it is unclear whether similar temperature sensitivity to floral reproduction also occurs in weed species. If pollen viability of weeds responds differently to temperature than that of crops, there would be consequences for seed production and competition.

Weed seed dispersal is also likely to be impacted by climate and CO_2. Many weed seeds are transported via water, and precipitation patterns are likely to change with global warming (IPCC, 2007). Wind disperses seeds of many weed species, including yellow salsify, spotted knapweed, leafy spurge, orange hawkweed, Canada thistle, perennial sowthistle, musk thistle, and others. Any increase in severe weather, particularly wind, is likely to aid in seed dispersal of these and other weeds. In addition, rising CO_2 levels may indirectly boost wind dispersal by increasing plant heights. Given the ability of some wind-dispersed invasive weeds to produce seed (e.g., tumbleweed can produce up to 25,000 seeds per plant), increases in weather extremes could have a significant effect on the rate and range of weed dispersal with species-specific effects on weed density and persistence within plant communities.

Unfortunately, few attempts have been made to quantify relative seed production of crops and weeds in response to climate change and/or CO_2. This may seem puzzling, since seed yield has been determined in so many crop species in response to climatic variables; however, it is mechanistically difficult to obtain seed production of weeds in situ given that seed shattering and temporal seed shedding is endemic to many weed species. Consequently, meta-analyses comparing the reproductive output of crop and "wild" species may only include a handful of common agronomic weeds (e.g., velvetleaf, sicklepod, barnyardgrass, and ladysthumb, see Appendix 1; Jablonski et al., 2002). In addition, comparisons of weed–crop seed production do not consider other means by which weed species may proliferate.

In contrast to most crop species, weeds may also reproduce asexually, that is, without fertilization or meiosis. Seed can also be apomictic, genetically identical to their parent (e.g., dandelion, orange hawkweed). Data for dandelion by McPeek and Wang (2007) indicated that elevated CO_2-grown plants produced 83% more inflorescences, with 32% more achenes. In addition, seed from elevated CO_2-grown dandelion were significantly heavier and had a higher germination percentage, leading to larger seedlings and earlier establishment in the subsequent generation. Asexual reproduction can also occur through vegetative propagation in which part of a plant becomes detached and develops into a separate individual. This may be of particular interest in weed biology since a number of studies have indicated that below ground growth (roots and rhizomes) can show disproportionate increases in response to elevated CO_2 (Bernsten and Woodward, 1992; Prior et al., 1994; Rogers et al., 1992). For example, significant increases in root:shoot ratio were observed in three years of field trials of Canada thistle grown at elevated CO_2 (Ziska et al. 2004a). In addition, both recent and projected increases in atmospheric CO_2 resulted in significant increases in root:shoot ratio for a number of pernicious weeds including field bindweed and perennial sowthistle (Figure 4.1) (Ziska, 2003a). Given that many perennial weeds have a strong below-ground root or rhizome system and can generate asexually through vegetative propagation, it is possible that

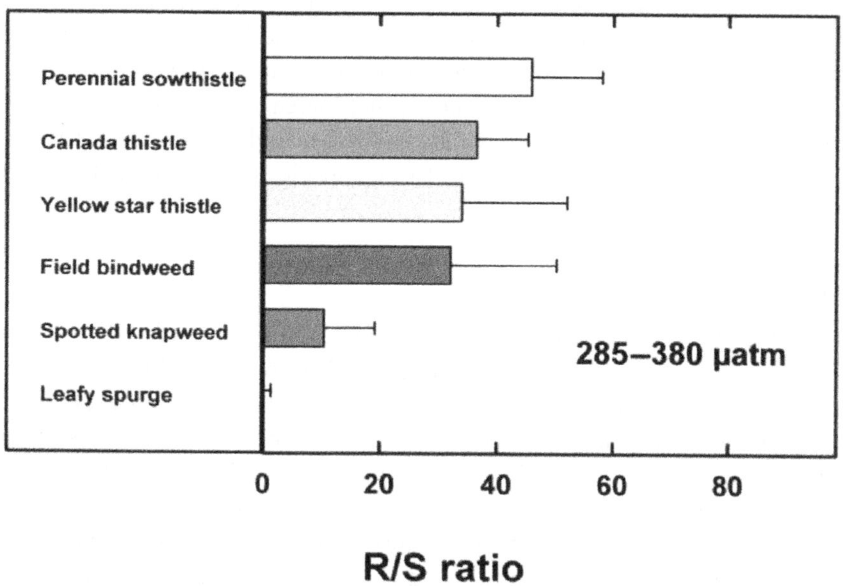

Figure 4.1 Change in allocation of additional carbon with increases in atmospheric carbon dioxide concentration during the twentieth century. Significant increases in root:shoot ratio were noted from four of the six invasive weed species examined, indicating potential changes in asexual reproduction with rising carbon dioxide.

substantial below-ground sinks could contribute to reproductive success of such weeds, providing a link between species establishment and carbon dioxide responsiveness.

Populations and Communities

Climate Change Migrations

How will climate change affect plant community composition, and weed distribution? Weed scientists have long recognized that temperature controls weed species success (Woodward and Williams, 1987). Thus, warming will impact the dissemination of weeds in both managed and unmanaged systems, with subsequent effects on their growth, reproduction, and competitive abilities. Such changes are reflected in recent plant hardiness zone maps demonstrating changes in plant distribution with a general increase in minimum winter temperatures (Figure 4.2).

Many of the most troublesome weeds in agriculture, both C_3 and C_4, are confined to tropical or subtropical areas (Holm et al., 1997) due to low temperature. High-latitude temperature limits of tropical species are set by accumulated degree days (Patterson et al., 1999), while low-latitude limits are determined, in part, by competitive ability at low temperatures (Woodward, 1988). Because many of the weeds associated with warm season crops originated in tropical or warm temperature areas, northward expansion of these weeds may accelerate with warming (Patterson, 1993; Rahman and Wardle, 1990). Other weeds in which the temperature sensitivity of growth has been examined with respect to range extension include cassia, showy crotalaria, cogongrass, Texas panicum, itchgrass, sesbania, and witchweed (Patterson, 1993, 1995). In addition to global warming resulting in likely latitudinal or altitudinal progression of weed species, it should be noted that increasing carbon dioxide alone could allow such range extensions. Preliminary data suggest that elevated CO_2 can increase plant tolerance to low temperatures in a number of weed species (Boese et al., 1997; Potvin and Strain, 1985; Sionit et al., 1981, 1987).

Weed Populations in Managed Systems

For maize and soybean crops within the United States, there is a clear latitudinal distinction between the great lakes (Michigan, Minnesota, Wisconsin) and gulf states (Alabama, Louisiana, Mississippi) with respect to weed species (Bunce and Ziska, 2000). Although there are no obvious differences in C_3 or C_4 species, or the ratio of annuals to perennials, the average crop losses (without herbicide use) are substantially larger for the gulf relative to the great lakes states for both maize (35% versus 22%) and soybeans (64% versus 22%). The greater crop losses in the southern gulf states are likely to be associated with a number of very aggressive weed species that are found in tropical or subtropical areas (e.g., prickly sida and Johnson grass). Warmer temperatures, in particular an increase in the number of frost-free days, may allow a northward expansion of these aggressive weeds into other areas of the Midwest, with subsequent impacts on maize and soybean production, or, to maintain production rates, greater herbicide use.

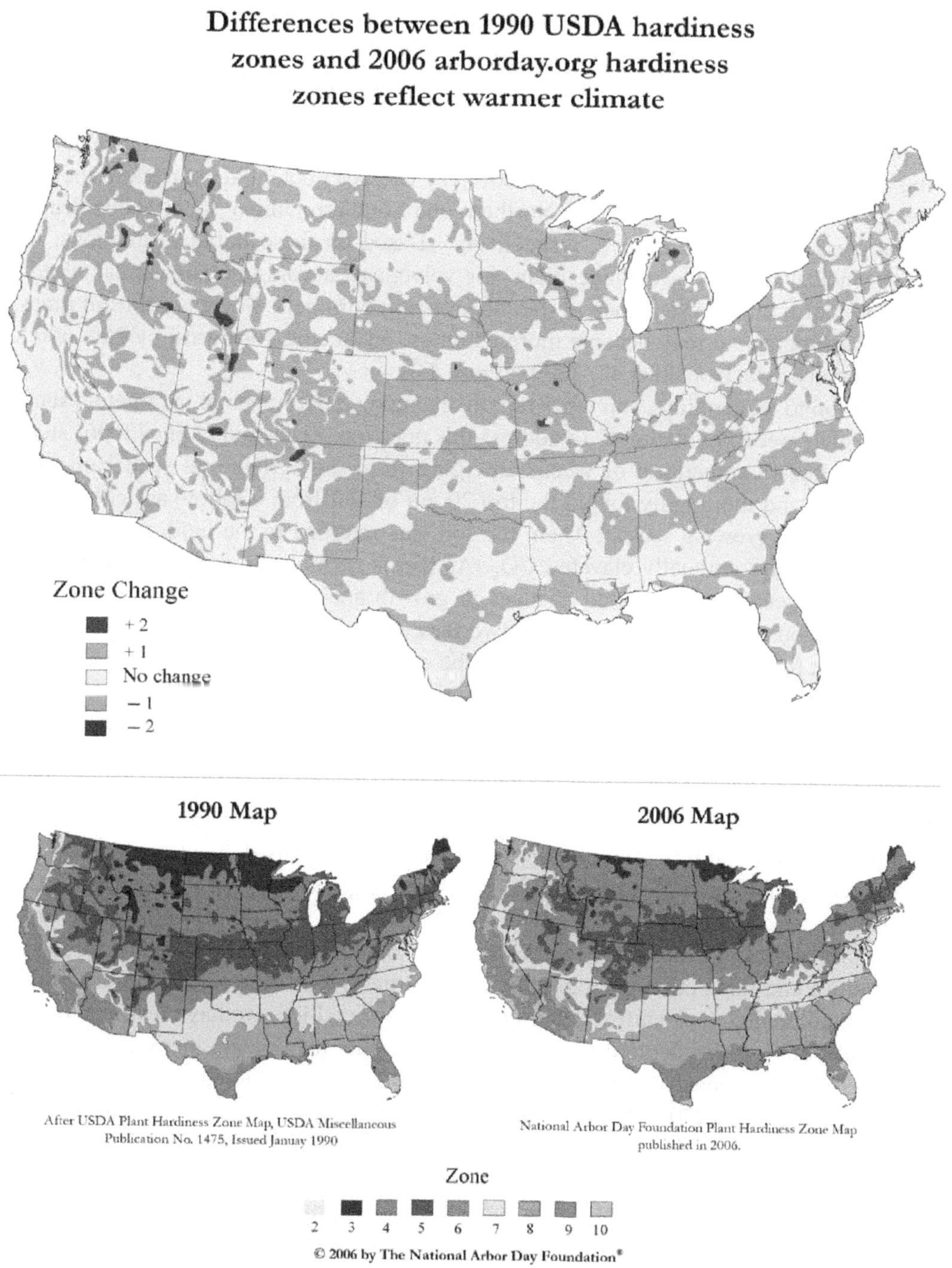

Figure 4.2 Change in hardiness zones from 1990 to 2006 as a function of warmer climate, principally warmer winters. Color version available online at http://www.arborday.org/media/map_change.cfm.

Biogeographic models that link species distributions and environmental factors (e.g., CLIMEX; Sutherst and Maywald, 1985; Sutherst et al., 2007) can be used to understand how cropping systems may be affected by shifts in geographic ranges of weeds. Using a "damage niche" hypothesis, where such a niche is characterized by the environmental conditions that make a weed abundant and competitive, a group of Cornell scientists has examined, using a "Business as Usual" climate scenario (IPCC, 2007) potential changes in the range of two weed species in maize for the Northern and Southern United States (velvetleaf and Johnsongrass, a C_3 and C_4 weed, respectively). Based on these initial evaluations, their analysis indicated that the damage niche for velvetleaf, a cold-tolerant annual weed, is likely to be of minor consequence in the mid-western corn belt; whereas Johnson grass, a warm-season perennial may become more common, advancing northward by 200–600 km by mid-century (MacDonald et al., 2009; Figure 4.3).

In addition to temperature, can other rapidly changing environmental parameters such as CO_2 also alter weed populations? To address this question, Ziska and Goins (2006) studied how CO_2 affects competition of weeds emerging from the agricultural seed bank. First, they removed and thoroughly mixed the top 20 cm of soil from a fallow agricultural field at the end of the growing season. Then, the soil was returned to the field, and exposed to either ambient or continually elevated CO_2. As plants germinated and grew from the soil the following year, they examined changes in growth and competition among C_3 and C_4 weed populations. The experiment ran for two seasons, and in the first, precipitation reached record levels, followed by near normal levels the next year. In the first year, only C_4 grasses were recruited, consistent with higher precipitation levels and the establishment of shallow rooted grasses associated with low O_2 soil conditions. In the second year, however, a broader range of weed species grew, and elevated CO_2 resulted in a greater incidence ($5\times$) number of C_3 weeds. This experiment emphasized the complexity of how multiple environmental changes can alter the relative distribution of C_3 and C_4 weeds within populations. Again, this reinforces the idea that weed populations are dynamic, not easily defined as either all C_3 or C_4, and will change not only with cultural practices (e.g., no-till), but with climate as well (Figure 4.4).

Overall, changes in temperature, carbon dioxide, and precipitation are likely to shift weed communities and population demographics around the world. However, there are limits to our ability to predict such shifts, for a number of reasons. First, we have few experimental data on this subject. Particularly lacking are studies that manipulate many environmental parameters simultaneously; such assessments are difficult methodologically even for large global regions (IPCC, 2007). Second, our weed data are out of date. It is difficult to project where weeds will be in the future, if you do not know where they are now. This relates to an earlier point regarding Holm's classification of "worst weeds" and the lack of knowledge regarding the dynamic nature of weed populations over time. As a consequence, the last national assessment of weed populations, done by David Bridges and the Weed Science Society of America, is 16 years old (Bridges, 1992). Yet, it is this "baseline" that is being used by the Cornell authors as well as many other agronomists to predict weed populations for crops with climatic change.

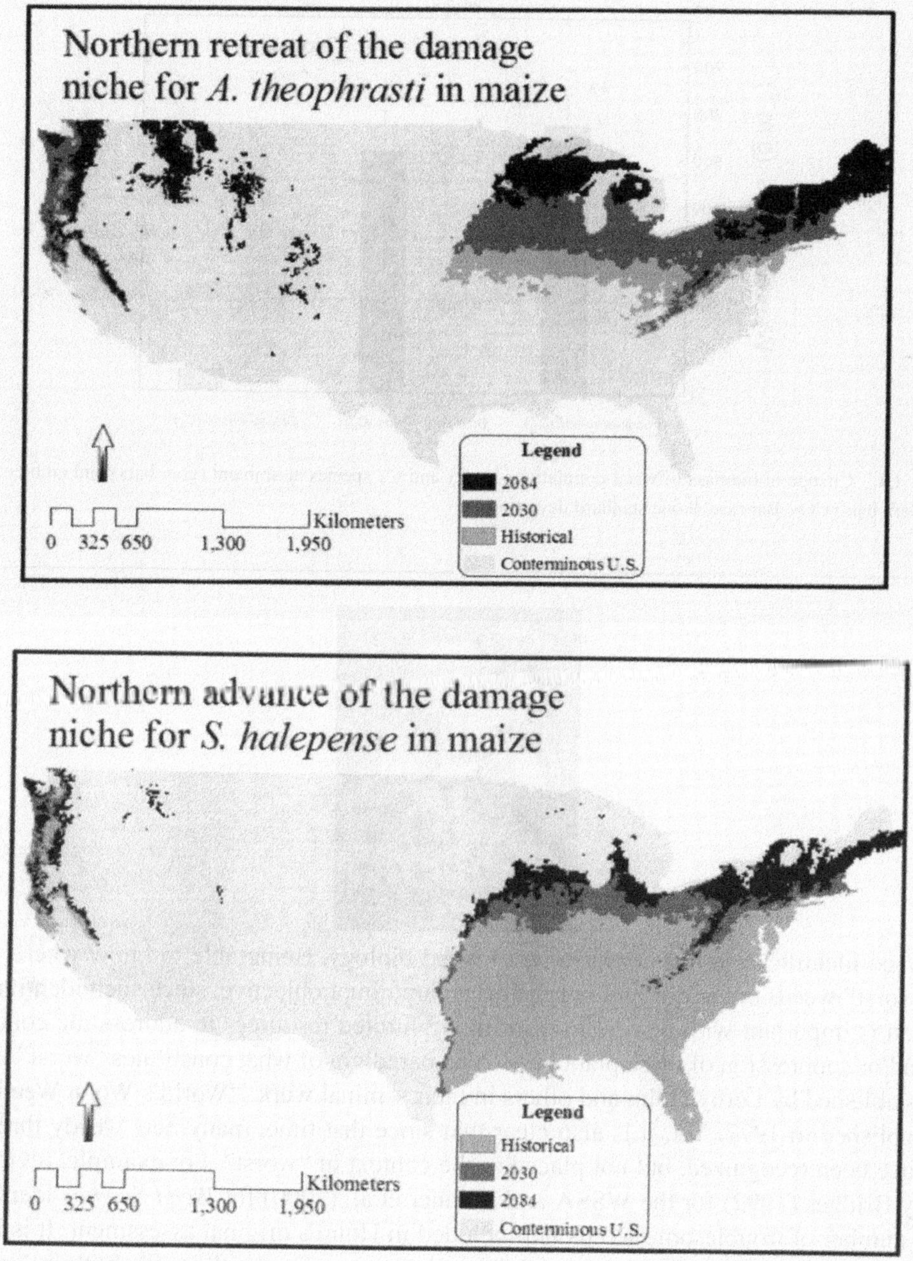

Figure 4.3 Change in the potential range of a C_3 weed, velvetleaf (*A. theophrasti*) and C_4 weed Johnsongrass (*S. halapense*). Note that with warming temperatures, the C_4 perennial weed may be more of a competitor with corn. (From MacDonald et al., 2009, *Agriculture, Ecosystems and Environment*, 130: 131–140.)

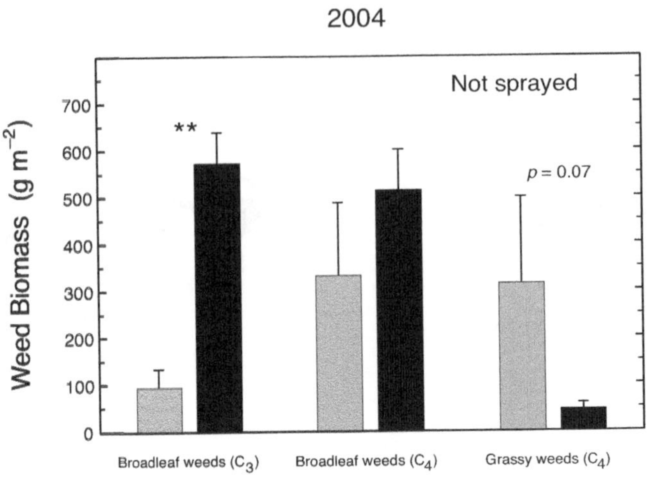

Figure 4.4 Change in biomass of weed populations by C_3 and C_4 species at ambient (grey bars) and ambient +250 ppm (dark bars) CO_2. Bars are ± one standard deviation.

Weed identification is a cornerstone of weed biology. Being able to know where the "worst" weeds are in our region remains a paramount objective, since such identification is important with respect to prioritizing limited resources to address the control and/or suppression of these plant pests. The paradigm of what constitutes "worst" was established by Leroy Holm and others in their seminal work: "World's Worst Weeds", published in 1977. Yet, it is also clear that since that time, many new weedy threats have been recognized, but not placed in the context of "worst". For example, reviews by Bridges (1992) for the WSSA and Skinner et al. (2000) for *Weed Science* identify a number of troublesome weeds not included in Holm's original assessment. It is important to compare and contrast the dynamic nature of "worst" weeds both in North America and globally to determine, in part, the cultural, biological, or environmental basis for how weed problems may wax and wane over time, and, discuss a preliminary basis by which we can begin to quantify what may constitute "worst" in the current weed scientist's lexicon.

Weed Populations in Unmanaged Systems

About 40% of Earth's land surface is used for grazing (Campbell et al., 1997). While these rangelands are usually water-limited and not suitable for long-term sustainable agriculture, they do support extensive livestock grazing (Polley et al., 2000).

A number of studies, therefore, have focused on how climate change, and the increase in atmospheric carbon dioxide in particular, could change plant community dynamics and rangeland quality. Such potential changes are of concern because for rangelands, the economic value of the land depends on the species composition of the plant community. There are a number of studies indicating that productivity of C_3 plants responds more to rising CO_2 than that of C_4 plants (Morgan et al., 2007; Polley et al., 1997; Poorter and Navas 2003). However, understanding the relative response of a plant species to carbon dioxide is not a simple reflection of photosynthetic pathway. As we have seen, CO_2 responses are complicated by temperature, water use, nutrient uptake, etc.; all of which complicate predictions of species dominance and plant community composition.

Secondary Succession and Ecosystems

Environmental changes will affect weed populations and communities in both managed and unmanaged plant systems. These changes, in turn, will alter ecosystem processes. Unfortunately, in examining the temporal and spatial variability of plant function over time (see Figure 3.1), we find that our lack of understanding is most obvious at extremes along the axes. That is to say, our knowledge of processes that are spatially small and occur quickly (e.g., gene expression), and those that are large in scale and occur slowly (e.g., ecosystem function) is minimal.

One essential ecological process related to ecosystem formation is succession. Following a disturbance, the plant species composition of an ecosystem changes over time until a stable community is achieved, a process known as succession. This stable community is termed the "climax community" and is often characterized as an ecosystem with a long-standing, relatively constant, species composition. Succession can be viewed as two components, primary and secondary. Primary succession is the establishment of plant species on land that has not been previously vegetated, such as lake-fill or rock that has been weathered to soil. Secondary succession in turn, is the re-establishment of plants after a serious environmental disturbance that has eliminated existing vegetation. The disturbance can be characterized as natural (e.g., hurricanes), or anthropogenic (e.g., abandoned farmland). Those plants which initially exploit the physical opportunities offered by succession possess many of the same characteristics as weed species, and are referred to as "pioneer" species in an ecological sense.

To examine the impact of projected climate changes on secondary succession for abandoned agricultural land, Ziska and colleagues (2004b) exploited a macro-environment of carbon dioxide and temperature that exists between large metropolitan centers (i.e., Baltimore) and surrounding rural areas (Western Maryland) (e.g., Berry and Colls 1990, Brazel et al. 2000). Over a five-year period, the urban site was 2.1°C warmer and had a CO_2 ~20% higher than the rural location, indicating that this gradient was a reasonable surrogate for changes in those variables projected to occur in this century (George et al., 2007, see B1 scenario, IPCC, 2007). Previous work had demonstrated that other abiotic variables measured across the same transect, including tropospheric ozone and nitrogen deposition, did not differ consistently (George et al., 2007).

To quantify macro-climatic differences associated with urbanization, Ziska and colleagues transplanted a common fallow farm soil with a common seed bank (about 35 different species, including C_3 and C_4, annuals and perennials) into urban, suburban, and rural locations along the transect. At each location, native soil was removed from four sites to establish four 2 m × 2 m holes to a depth of 110 cm. The holes were then filled with the fallow farm soil (Figure 4.5).

In the first year of the experiment, shoots of pioneer/weedy species (>90% lambsquarters) grew 2–3 times more in the urban site than in the rural site. However, by the end of 5 years, warmer temperatures, longer growing seasons (greater number of frost-free days), and higher CO_2 associated with the urban side of the transect were correlated with increased biomass

Figure 4.5 Change in biomass of annual weeds following a disturbance in a rural farm site (on left) and urban site (on right). Soil was from a common location and contained the same seed bank. Macro-climatic differences particularly warmer temperatures, longer growing season, and more CO_2 were associated with the urban location. Additional details are given in George et al. (2007) and Ziska et al. (2003).

production by both herbaceous and woody perennial species. In addition, by the end of the study, species demography of plant communities differed among sites, with a greater ratio of woody and herbaceous perennials to annuals in the urban location than the rural location (George et al., 2009). Interestingly, many of the perennials which came to dominate the urban site were fast-growing woody species. The basis for the faster transition from annual to perennial species is unclear. Greater first-year biomass and litter accumulation at the urban site may have suppressed the subsequent seed germination of annual species, accelerating changes in species composition (Figure 4.6).

If urban macro-environments reflect future global change conditions, these data suggest a faster rate of secondary succession in a warmer, higher CO_2 world. Such an increase would accelerate demographic change, and could change weed species dominance and ecosystem function. Bazzaz (1996) has also suggested that, because of the relatively fast rate of global change, plant species that require a long establishment time may be excluded from many locations. Rather, he suggests that fast-growing perennials may tend to dominate ecosystems in an elevated CO_2 environment, suggestions consistent with the results of the urban–rural succession study (George et al., 2009).

Evolution in Plant Populations with Climate Change

Given the biological consequences of CO_2 and climate change, what role might they play in the evolution of weeds and crops?

Some may argue that climate change will happen too fast for evolutionary responses to occur (i.e., species may either acclimate or become extinct). However, while this may be true for those species with long generation times and small population sizes (e.g., trees and large mammals), in which evolutionary change may occur on time scales of millennia, it may not be true for weeds. For example, since the introduction of herbicides, herbicide pressure has selected for weed biotypes with resistance, evidence that evolutionary responses to environmental perturbations can be rapid in plant species that impact human society. In that context, it is perhaps more pragmatic to calculate the biological circumstances associated with climate/CO_2 that would allow for evolution to occur quickly. We would argue that rapid evolutionary change is more likely in populations with high fecundity and genetic diversity, with short generation times and high intrinsic rates of growth. Weeds (and some crops) have all of these characteristics.

An elegant experiment by Franks et al. (2007) examined the effect of one likely aspect of climate change—drought—on the evolutionary response of wild mustard, a weed that occurs throughout North America. Ancestral genotypes that had not been subjected to drought were raised concurrently with descendent (postdrought) genotypes and ancestor × descendant hybrids. They reported that descendents flowered earlier than ancestors, with ancestor × descendent times being intermediate. This study supports the idea that imposition of drought selected for early flowering, that flowering times were heritable, and that selection intensities were sufficient to account for the observed evolutionary change. Natural selection for drought

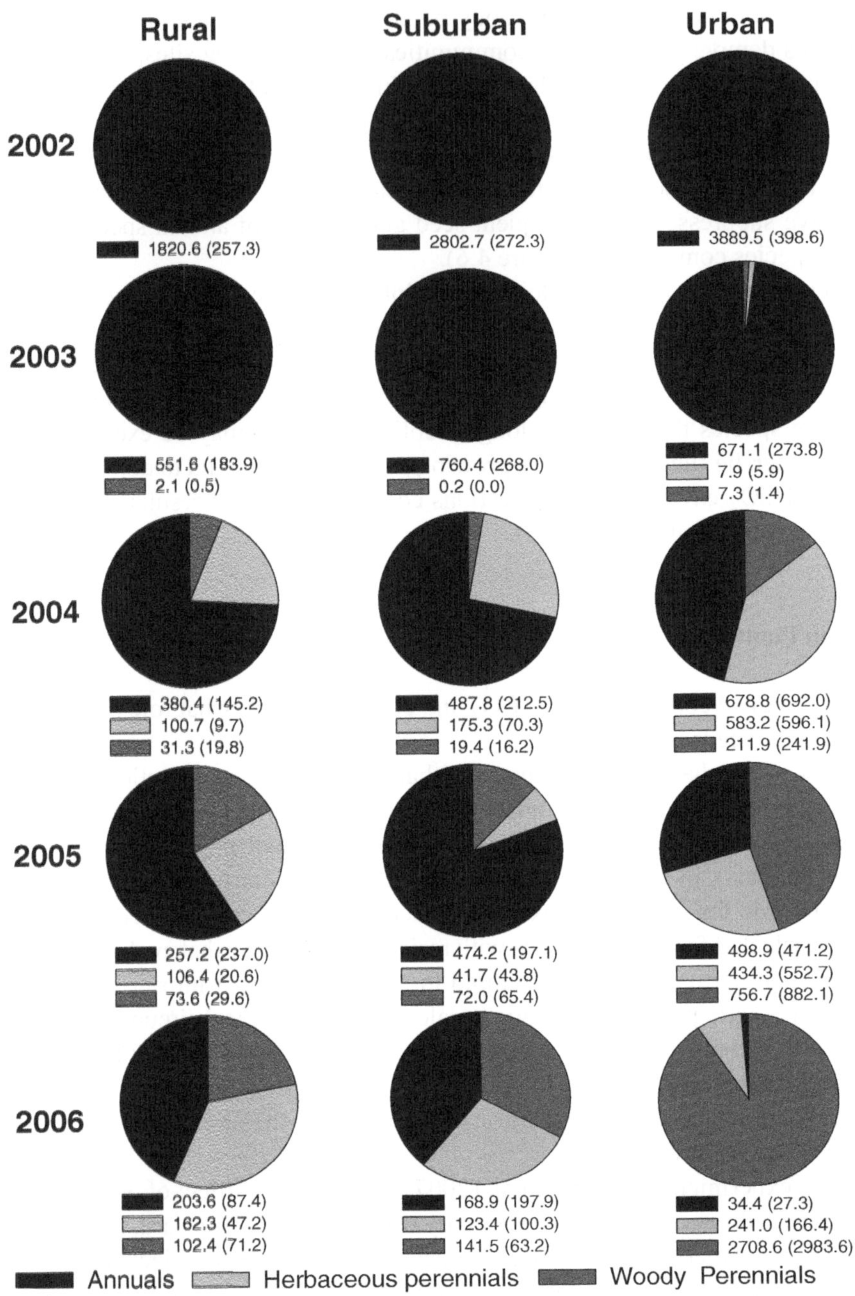

Figure 4.6 Change in plant demographics over a 5-year period during secondary succession. The urban location simulated environmental conditions of temperature and CO_2 anticipated in the IPCC B1 scenario (George et al., 2009).

Table 4.3 Significance of equality of variance for "old" and "new" oat cultivars for select vegetative and growth characteristics averaged over all CO_2 concentrations.

Variable	"Old"	"New"
Leaf area	*	
Leaf wt.	**	
Tiller wt.	**	
Weight tiller^{-1}	***	
Tiller no.	*	
Root wt.	*	
Total wt.	***	
RGR	**	

* $P < 0.05$; ** $P < 0.01$; *** $P < 0.001$.
An asterisk indicates a greater degree of phenotypic variation. Root:shoot ratio, LAR and NAR were not significant.

escape resulted in adaptive evolution in this species in just a few generations. If natural selection can begin to adapt weed species such as wild mustard to climatic extremes, or to atmospheric carbon dioxide, can similar adaptation be anticipated among crop species? Have crops already adapted to climate change and/or increasing CO_2?

A study on cultivated oat lines released form the 1920s and the 1990s from each of seven different breeding locations, indicated that newer lines had not intrinsically adapted to the change in CO_2 concentration that occurred during that time period (ca. 80 ppm CO_2 increase; Ziska and Blumenthal 2007). Rather, the data suggested that the older oat lines (released during the 1920s) had the stronger response to CO_2. Overall, the older lines had greater phenotypic variability for all of the parameters tested, while the modern (1990s) lines were much more uniform (Table 4.3).

Certainly at the agronomic level, artificial and natural selections for desired traits are not synergistic. Given this, what are the likely responses of cultivated and wild, weedy types to a similar environmental change? To determine whether recent or projected increases in CO_2 favor cultivated or weedy (red) rice, Ziska and McClung (2008) assessed the initial growth and vegetative characteristics of six wild type and six cultivated rice lines grown at carbon dioxide concentrations that corresponded roughly to the 1940s, the current CO_2, and the CO_2 projected for the middle of this century (300, 400, and 500 ppm, respectively). Results from this study indicate that recent or projected changes in atmospheric carbon dioxide favored red rice over cultivated rice, both *Oryza sativa*, when grown 100 and 200 ppm above the 300 ppm CO_2 baseline (Ziska and McClung 2008). Overall, photosynthesis, leaf area, and related characteristics or red rice responded more strongly than those of cultivated rice (Figure 4.7).

Can artificial selection for the development of new varieties keep pace with natural selection that occurs in indigenous wild (red) rice populations as atmospheric carbon dioxide increases (or with greater droughts, or greater occurrence of extreme temperatures, etc.)? It

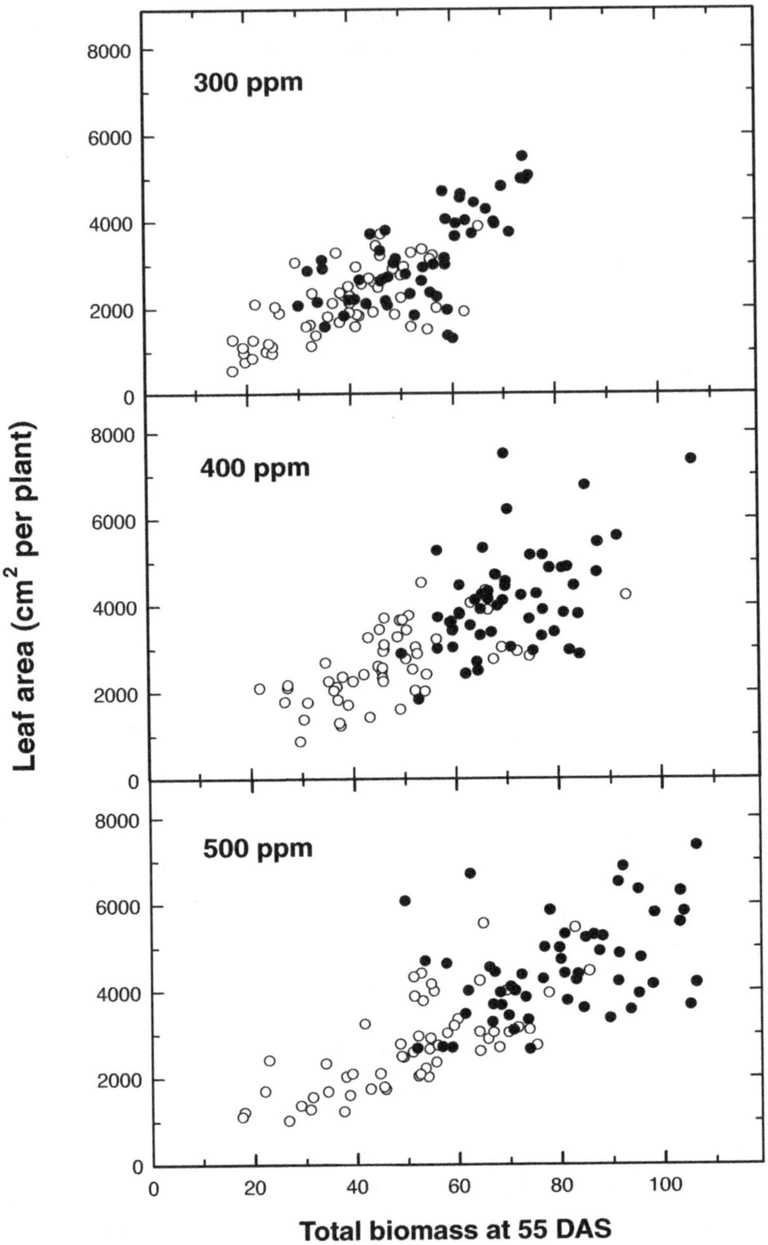

Figure 4.7 Change in leaf area as a function of total plant biomass 55 days after sowing (panicle initiation stage of growth) for six cultivated rice lines (CR, open circles) and six red rice lines (weedy biotypes, closed circles) as a function of CO_2 concentration. The CO_2 concentrations correspond roughly to the beginning and end of the twentieth century (300 and 400 ppm, respectively); and that projected for the middle of the twenty-first century (500 ppm). Note that with increasing CO_2, it is the weedy rice biotypes that show the greatest response.

can be argued that natural selection is slower than artificial selection since artificial selection can restrict the offspring of subsequent generations to a limited set of desired criteria (i.e., rarely is natural selection an all-or-none phenomenon). However, as with the Brassica experiment, evolution of phenological traits (e.g., flowering), can be rapid for some weedy species in response to climatic change (e.g., drought; Franks et al., 2007).

Summary

In the last two chapters, we have examined how climate and rising carbon dioxide are likely to affect various aspects of weed biology, from the cellular level to the ecosystem level. This chapter has focused on how a changing climate might influence competitive interactions between and/or within weeds and crops. We have provided examples of how climate/CO_2 can modify competition; weed populations and communities, particularly with respect to where weeds might grow in the future; ecosystem responses, using urbanization as a potential surrogate for future climates, and how evolution could act differently in weeds and crops. By necessity, we have emphasized managed agricultural ecosystems (e.g., croplands and rangelands), because this is where the bulk of the experimental data are available. In the following chapters, we examine how these CO_2–climatic changes in weed biology will alter the damage inflicted by weeds in a number of human and natural systems.

References

Alm DM, McGiffen ME, Jr. Hesketh JD 1991. Weed phenology. In: Hodges T (ed) *Predicting Crop Phenology*. CRC Press, Boca Raton, FL, pp. 191–218.

Bazzaz FA. 1996. *Plants in Changing Environments: Linking Physiological, Population and Community Ecology*. Cambridge University Press, Cambridge, UK, 320 pp.

Bazzaz FA, Carlson RW. 1984. The response of plants to elevated CO_2. I. Competition among an assemblage of annuals at different levels of soil moisture. *Oecologia* **62**:196–198.

Bazzaz FA, Garbutt K. 1988. The response of annuals in competitive neighborhoods: effects of elevated CO_2. *Ecology* **69**:937–946.

Bazzaz FA, Garbutt K, Reekie ER, Williams WE. 1989. Using growth analysis to interpret competition between a C_3 and a C_4 annual under ambient and elevated CO_2. *Oecologia* **79**:223–235.

Belote RT, Weltzin JF, Norby RJ. 2003. Response of an understory plant community to elevated CO_2 depends on differential responses of dominant invasive species and is mediated by soil water availability. *New Phytol.* **161**:827–835.

Bernsten GN, Woodward FI. 1992. The root system architecture and development of *Senecio vulgaris* in elevated CO_2 and drought. *Funct. Ecol.* **6**:324–333.

Berry RD, Colls JJ. 1990. Atmospheric carbon dioxide and sulphur dioxide on an urban/rural transect-I. Continuous measurements at the transect ends. *Atmos. Environ.* **24**:2681–2688.

Boese, SR, Wolfe DW, Melkonian J. 1997. Elevated CO_2 mitigates chilling-induced water stress and phosynthetic reduction during chilling. *Plant Cell Environ.* **20**:625–632.

Brazel A, Selover N, Vose R, Heisler G. 2000. The tale of two climates—Baltimore and Phoenix urban LTER sites. *Climate Res.* **15**:123–135.

Bridges DC. 1992. *Crop Losses Due to Weeds in the United States*. Weed Science Society of America, Champaign, IL, p. 403.

Bunce JA. 2001. Are annual plants adapted to the current concentration of carbon dioxide? *Int. J Plant Sci.* **162**:1261–1266.

Bunce JA, Ziska LH. 2000. Crop ecosystem responses to climatic change: crop/weed interactions. In: Reddy KR, Hodges HF (eds) *Climate Change and Global Crop Productivity*. CABI Publishing, New York, pp. 333–348.

Campbell BD, Smith DMS, McKeon GM. 1997. Elevated CO_2 and water supply interactions in grasslands: a pastures and rangelands management perspective. *Global Change Biol.* **3**:177–187.

Chen D-X, Coughenour MB, Eberts D, Thullen J. 1994. Interactive effects of CO_2 enrichment and temperature on the growth of dioecious *Hydrilla verticillata*. *Env. Exp. Bot.* **34**:345–353.

Condon MA, Sasek TW, Strain BR. 1992. Allocation patterns in two tropical vines in response to elevated atmospheric CO_2. *Funct. Ecol.* **6**:680–685.

Dillenburg LR, Teramura AH, Forseth IN, Whigham DF. 1995. Photosynthetic and biomass allocation responses of *Liquidambar styraciflua* (Hamamelidaceae) to vine competition. *Am. J. Bot.* **82**:454–461.

Dippery JK, Tissue DT, Thomas RB, Strain BR. 1995. Effects of low and elevated CO_2 on C_3 and C_4 annuals. *Oecologia* **101**:13–20.

Dukes JS. 2002. Comparison of the effect of elevated CO_2 on an invasive species (*Centaurea solstitialis*) in monoculture and community settings. *Plant Ecol.* **160**:225–234.

Esashi Y, Hase S, Kojima, K. 1987. Light actions in the germination of cocklebur seeds: V. Effects of ethylene, carbon dioxide, oxygen on germination in relation to light. *J. Exp. Bot.* **38**:702–710.

Firbank LG, Watkinson AR. 1985. On the analysis of competition with two-species mixtures of plants. *J. Appl. Ecol.* **22**:503–517.

Fitter AH, Fitter RSR. 2002. Rapid changes in flowering time in British plants. *Science* **296**:1689–1691.

Franks SJ, Sim S, Weis AE. 2007. Rapid evolution of flowering time by an annual plant in response to a climate fluctuation. *Proc. Natl. Acad. Sci. U. S. A.* **104**:1278–1282.

Freeden AL, Field CB. 1995. Contrasting leaf and ecosystem CO_2 and H_2O exchange in *Avena fatua* monoculture: growth at ambient and elevated CO_2. *Photo. Res.* **43**:263–271.

Garbutt K, Williams WE, Bazzaz FA. 1990. Analysis of the differential response of five annuals to elevated carbon dioxide during growth. *Ecology* **71**:1185–1194.

George K, Ziska LH, Bunce JA, Quebedeaux B. 2007. Elevated atmospheric CO_2 concentration and temperature across an urban–rural transect. *Atmos. Environ.* **41**:7654–7665.

George K, Ziska LH, Bunce JA, Quebedeaux B, Hom JL, Wolf J, Teasdale JR. 2009. Macroclimate associated with urbanization increases the rate of secondary succession from fallow soil. *Oecologia* **159**:637–647.

Gordo O, Sanz JJ. 2005. Phenology and climate change: a long-term study in a Mediterranean locality. *Oecologia* **146**:484–495.

Hatfield JL, Boote KJ, Fay P, Hahn L, Izaurralde C, Kimball BA, Mader T, Morgan JA, Ort D, Polley W, Thomson A, Wolfe D. 2008. Agriculture. In: *The Effects of Climate Change on Agriculture, Land Resources, Water Resources, and Biodiversity in the United States*. A report by the U.S. Climate Change Science Program and the Subcommittee on Global Change Research. Washington, DC, pp. 21–74.

Heichel GH, Jaynes RA. 1974. Stimulating emergence and growth of *Kalmia* genotypes with carbon dioxide. *Hort. Sci.* **9**:60–62.

Hirose T, Ackerly DD, Traw MB. 1996. Effects of CO_2 elevation on canopy development in the stands of two co-occurring annuals, *Oecologia* **108**:215–223.

Holm LG, Doll J, Holm E, Pancho J, Herverger J. 1997. *Worlds Weeds: Natural Histories and Distribution*. John Wiley & Sons, New York, 1129 pp.

Holm LG, Plucknett DL, Pancho JV, Herberger JP. 1977. *The Worlds Worst Weeds. Distribution and Biology*. University of Hawaii Press, Honolulu, HI, 609 pp.

Huxman TE, Hamerlynck EP, Smith SD. 1999. Reproductive allocation and seed production in *Bromus madritensis* ssp. *rubens* at elevated atmospheric CO_2. *Funct Ecol* **13**:769–777.

IPCC: Climate Change. 2007. *Impacts, Adaptation and Vulnerability*. IPCC Secretariat, Geneva, Switzerland, 986 pp.

Jablonski LM, Wang X, Curtis PS. 2002. Plant reproduction under elevated CO_2 conditions: a meta-analysis of reports on 79 crop and wild species. *N. Phytol.* **156**:9–26.

Kropff MJ, Spitters CJT. 1991. A simple model of crop loss by weed competition from early observations on relative leaf area of the weeds. *Weed Res.* **31**:97–105.

Laurance WF, Delamonica P, D'Angelo S, Jerozolinski A, Pohl L. 2001. Rain forest fragmentation and the structure of Amazonian liana communities. *Ecology* **82**:105–116.

MacDonald AJ, Riha SJ, DiTommaso A, DeGaetano AT. 2009. Climate change and the geography of weed damage: analysis of U.S. maize systems suggests the potential for significant range transformations. *Agric. Ecosyst. Enviorn.* **130**:131–140.

McPeek TM, Wang X. 2007. Reproduction of dandelion (*Taraxacum officinale*) in a higher CO_2 environment. *Weed Sci.* **55**:334–340.

Mohan JE, Clark JS, Schlesinger WH. 2004. Genetic variation in germination, growth and survivorship of red maple in response to sub-ambient through elevated atmospheric CO_2. *Global Change Biol.* **10**:233–247.

Morgan JA, Milchunas DG, LeCain DR, West M, Mosier AR. 2007. Carbon dioxide enrichment alters plant community structure and accelerates shrub growth in the shortgrass steppe. *Proc. Natl. Acad. Sci. U. S. A.* **104**:14724–14729.

Morose SR, Bazzaz FA. 1994. Elevated CO_2 and temperature alter recruitment and size hierarchies in C_3 and C_4 annuals. *Ecology* **75**:966–975.

Parmesan C, Ryrholm N, Stefanescu C, Hill JK, Thomas CD, Descimon H, Huntley B, Kaila L, Kullberg J, Tammaru T, Tennent J, Thomas JA, Warren M. 1999. Poleward shift of butterfly species' ranges associated with regional warming. *Nature* **399**:579–583.

Patterson DT. 1986. Responses of soybean (*Glycine max*) and three C_4 grass weeds to CO_2 enrichment during drought. *Weed Sci.* **34**:203–210.

Patterson DT. 1993. Implications of global climate change for impacts of weeds, insects and plant diseases. *Int. Crop Sci.* **1**:273–280.

Patterson DT. 1995. Effects of environmental stress on weed/crop interactions. *Weed Sci.* **43**:483–490.

Patterson DT, Flint EP. 1980. Potential effects of global atmospheric CO_2 enrichment on the growth and competitiveness of C_3 and C_4 weed and crop plants. *Weed Sci.* **28**:71–75.

Patterson DT, Meyer CR, Flint EP, Quimby PC Jr. 1979. Temperature responses and potential distribution of itchgrass (*Rottboellia exaltata*) in the United States. *Weed Sci.* **27**:77–82.

Patterson DT, Westbrook JK, Joyce RJV, Rogasik J. 1999. Weeds, insects and diseases. *Clim. Change* **43**:711–727.

Phillips OL, Martinez RV, Arroyo L, Baker TR, Killeen T, Lewis SL, Malhi Y, et al. 2002. Increasing dominance of large lianas in Amazonian forests. *Nature* **418**:770–774.

Polley HW, Johnson HB, Mayeux HS. 1994. Increasing CO_2: Comparative responses of the C_4 grass *Schizachyrium* and grassland invader *Prosopis*. *Ecol.* **75**:976–988.

Polley HW, Mayeux HS, Johnson HV, Tischler CR. 1997. Viewpoint: atmospheric CO_2, soil water, and shrub/grass ratios on rangelands. *J. Range Manag.* **50**:278–284.

Polley HW, Morgan JA, Campbell BD, Smith MS. 2000. Crop ecosystem responses to climatic change: rangelands. In: Reddy KR, Hodges HF (eds) *Climate Change and Global Crop Productivity*. CABI Publishing, New York, pp. 293–314.

Polley HW, Johnson HB, Tischler CR. 2002. Woody invasion of grasslands: evidence that CO_2 enrichment indirectly promotes establishment of *Prosopis glandulosa*. *Plant Ecol.* **164**:85–94.

Poorter H. 1993. Interspecific variation in the growth response of plants to an elevated ambient CO_2 concentration, *Vegetation* **104/105**:77–97.

Poorter H, Navas M-L. 2003. Plant growth and competition at elevated CO_2: on winners, losers and functional groups. *N. Phytol.* **157**:175–198.

Potvin C, Strain BR. 1985. Effects of CO_2 enrichment and temperature on growth in two C_4 weeds, *Echinochloa crus-galli*, and *Eleusine indica*. *Can. J. Bot.* **63**:1495–1499.

Prior SA, Rogers HH, Runion GB, Mauney JR. 1994. Effects of free-air enrichment on cotton root growth. *Agric. Forest Met.* **70**:69–86.

Putz FE, Mooney HA. 1991. In: Putz FE, Mooney HA (eds) *The Biology of Vines*. Cambridge University Press, Cambridge, UK, 542 pp.

Rahman A, Wardle DA. 1990. Effects of climate change on cropping weeds in New Zealand. In: Prestridge RA, Pottinger PP (eds) *The Impact of Climate Change on Pests, Diseases, Weeds and Beneficial Organisms Present in New Zealand Agricultural and Horticultural Systems*. New Zealand Ministry for the Environment, Wellington, NZ, pp. 107–112.

Reich PB, Hobbie SE, Lee T, Ellsworth DS, West JB, Tilman D, Knops JMH, Naeem S, Trost J. 2006. Nitrogen limitation constrains sustainability of ecosystem response to CO_2. *Nature* **440**:922–925.

Rogers HH, Peterson CM, McCrimmon JN, Cure JD. 1992. Response of plant roots to elevated atmospheric carbon dioxide. *Plant Cell Environ.* **15**:749–752.

Sasek TW, Strain BR. 1988. Effects of carbon dioxide enrichment on the growth and morphology of Kudzu (*Pueraria lobata*). *Weed Sci.* **36**:28–36.

Sasek TW, Strain BR. 1989. Effects of carbon dioxide enrichment on the expansion and size of kudzu (*Pueraria lobata*) leaves. *Weed Sci.* **37**:23–28.

Sasek TW, Strain BR. 1990. Implications of atmospheric CO_2 enrichment and climatic change for the geographical distribution of two introduced vines in the USA. *Clim. Change* **16**:31–51.

Sasek TW, Strain BR. 1991. Effects of carbon dioxide enrichment on the growth and morphology of an introduced and a native honeysuckle vine species. *Am. J Bot.* **78**:69–75.

Schnitzer SA, Bongers F. 2002. The ecology of lianas and their role in forests. *TREE* **17**:223–230.

Singer BD, Ziska LH, Frenz DA, Gebhard DE, Straka, JG. 2005. Increasing Amb a 1 content in common ragweed (*Ambrosia artemisiifolia*) pollen as a function of rising atmospheric CO_2 concentration. *Funct. Plant Biol.* **32**:667–670.

Sionit N, Strain BR, Beckford HA. 1981. Environmental controls on the growth and yield of okra. I. Effects of temperature and of CO_2 enrichment at cool temperature. *Crop Sci.* **21**:885–888.

Sionit N, Strain BR, Flint EP. 1987. Interaction of temperature and CO_2 enrichment on soybean: photosynthesis and seed yield. *J. Plant Sci.* **67**:629–636.

Skinner K, Smith L, Rice P. 2000. Using noxious weed lists to prioritize targets for developing weed management strategies. *Weed Sci.* **48**:640–644.

Smith SD, Strain BR, Sharkey TD. 1987. Effects of CO_2 enrichment on four Great Basin grasses. *Funct. Ecol.* **1**:139–143.

Smith SD, Huxman TE, Zitzer SF, Charlet TN, Housman DC, Coleman JS, Fenstermaker LK, Seemann JR, Nowak RS. 2000. Elevated CO_2 increases productivity and invasive species success in an arid ecosystem. *Nature* **408**:79–82.

Steinmaus SJ, Prather TS, Holt JS. 2000. Estimation of base temperature for nine weed species. *J. Exp. Bot.* **51**:275–286.

Sutherst RW, Maywald GF. 1985. A computerised system for matching climates in ecology. *Agri. Eco. Environ.* **13**:281–299.

Sutherst RW, Baker RHA, Coakley SM, Harrington R, Kriticos DJ, Scherm H. 2007. Pests under global change—meeting your future landlords? In: Canadell JC, Pataki DE, Pitelka LF (eds) *Terrestrial Ecosystems in a Changing World*, Springer, Berlin, Heidelberg, pp. 211–226.

Thelen GC, Vivanco JM, Newingham B, Good W, Bais HP, Landres P, Caesar A, Callaway RM. 2005. Insect herbivory stimulates allelopathic exudation by an invasive plant and the suppression of natives. *Ecol. Lett.* **8**:209–217.

Tremmel DC, Patterson DT. 1993. Responses of soybean and five weeds to CO_2 enrichment under two temperature regimes. *Can. J. Plant Sci.* **73**:1249–1260.

Vengris J, Colby WG, Drake M. 1955. Plant nutrient competition between weeds and corn. *Agron. J.* **47**:213–216.

Wan S, Yuan T, Bowdish S, Wallace L, Russell SD, Luo Y. 2002. Response of an allergenic species, *Ambrosia psilostachya* (Asteraceae), to experimental warming and clipping: implications for public health. *Am. J. Bot.* **89**:1843–1846.

Wayne PM, Carnelli AL, Connolly J, Bazzaz FA. 1999. The density dependence of plant responses to elevated CO_2. *J. Ecol.* **87**:183–192.

Wayne P, Foster S, Connolly J, Bazzaz F, Epstein P. 2002. Production of allergenic pollen by ragweed (*Ambrosia artemisiifolia* L.) is increased in CO_2-enriched atmospheres. *Ann. Allergy Asthma Immunol.* **88**:279–282.

Woodward FI. 1988. Temperature and the distribution of plant species. In: Long SP, Woodward FI (eds) *Plants and Temperature*. University of Cambridge, Cambridge, UK, pp. 59–75.

Woodward FI, Williams BG. 1987. Climate and plant distribution at global and local scales. *Vegetatio* **69**:189–197.

Zangerl AR, Bazzaz FA. 1984. The response of plants to elevated CO_2. II. Competitive interactions among annual plants under varying light and nutrients. *Oecologia* **62**:412–417.

Zavala JA, Casteel CL, DeLucia EH, Berenbaum MR. 2008. Anthropogenic increase in carbon dioxide compromises plant defense against invasive insects. *Proc. Natl. Acad. Sci. USA* **105**:5129–5133.

Ziska LH. 2002. Influence of rising atmospheric CO_2 since 1900 on early growth and photosynthetic response of a noxious invasive weed, Canada thistle (*Cirsium arvense*). *Funct. Plant Biol.* **29**:1387–1392.

Ziska LH. 2003a. Evaluation of the growth response of six invasive species to past, present and future carbon dioxide concentrations. *J Exp. Bot.* **54**:395–404.

Ziska LH. 2003b. The impact of nitrogen supply on the potential response of a noxious, invasive weed, Canada thistle (*Cirsium arvense*) to recent increases in atmospheric carbon dioxide. *Physiol. Plant* **119**:105–112.

Ziska LH, Blumenthal DM. 2007. Empirical selection of cultivated oat in response to rising atmospheric carbon dioxide. *Crop Sci.* **47**:1547–1552.

Ziska LH, Bunce JA. 1993. The influence of elevated CO_2 and temperature on seed germination and soil emergence. *Field Crops Res.* **34**:147–157.

Ziska LH, Bunce JA, Goins EW. 2004b. Characterization of an urban–rural CO_2/temperature gradient and associated changes in initial plant productivity during secondary succession. *Oecologia* **139**:454–458.

Ziska LH, Caulfield FA. 2000. Rising carbon dioxide and pollen production of common ragweed, a known allergy inducing species: implications for public health. *Aust. J. Plant Physiol.* **27**:893–898.

Ziska LH, Faulkner SS, Lydon J. 2004a. Changes in biomass and root:shoot ratio of field-grown Canada thistle (*Cirsium arvense*), a noxious, invasive weed, with elevated CO_2: implications for control with glyphosate. *Weed Sci.* **52**:584–588.

Ziska LH, Gebhard DE, Frenz DA, Faulkner S, Singer BD. 2003. Cities as harbingers of climate change: common ragweed, urbanization, and public health. *J. Allergy Clin. Immun.* **111**:290–295.

Ziska LH, George K. 2004. Rising carbon dioxide and invasive, noxious plants: potential threats and consequences. *World Res. Rev.* **16**:427–447.

Ziska LH, Goins EW. 2006. Elevated atmospheric carbon dioxide and weed populations in glyphosate treated soybean. *Crop Sci.* **46**:1354–1359.

Ziska LH, McClung A. 2008. Differential response of cultivated and weedy (red) rice to recent and projected increases in atmospheric carbon dioxide. *Agron J.* **100**:1259–1263.

Ziska LH, Teasdale JR. 2000. Sustained growth and increased tolerance to glyphosate observed in a C_3 perennial weed, quackgrass (*Elytrigia repens*), grown at elevated carbon dioxide. *Austral. J Plant Physiol.* **27**:159–166.

Ziska LH, Reeves JB III, Blank R. 2005. The impact of recent increases in atmospheric CO_2 on biomass production and vegetative retention of Cheatgrass (*Bromus tectorum*): implications for fire disturbance. *Global Change Biol.* **11**:1325–1332.

5 Weeds on the Farm: Assessing the Role of Climate Change and CO_2 on Agricultural Productivity

"The philosopher who said that work well done never needs doing over . . . never weeded a garden."

Ray D. Everson

Introduction

In this, and the succeeding chapters, we will begin to examine in more detail the damage done by weeds, and how rising CO_2 and climate change may modify that damage. For agronomists, the objectionable nature of weeds arises from their negative impact on food production systems; either through direct competition with crops for water, light, nutrients, etc.; or indirectly by reducing production quality, increasing harvest interference or acting as hosts for other pest vectors such as insects or disease. Agriculture, in its simplest sense, represents a managed community that consists of the crop (desired plant species) and weeds (undesired plant species). Weeds, since the inception of agriculture, have reduced crop productivity and food security. According to Oerke (2006), weeds cause much higher potential crop losses (34%) than insect pests or pathogens (losses of 18% and 16%, respectively). The actual impact of weeds on global food production is frequently overlooked due to their consistent presence in agro-ecosystems. While understanding the risk to crop yield and quality, farmers have become inured to the presence of weeds, paying more attention to unknown, unpredictable epidemics of insect pests or pathogens. In 1991, the estimated average annual monetary loss caused by weeds with current control strategies in 46 crops in the United States was $4.1 billion. If herbicides were not used, this loss was estimated to be $19.6 billion. Field crop losses accounted for 82% of this total (Bridges, 1992). The damage done by weeds in agriculture has been (and likely will continue to be) the principal focus of most weed science research done by government, university, and private industry, and deservedly so.

Crop Losses and Weeds: An Overview

As we explored in the previous chapters, there are a number of ways by which the environment affects weed–crop competition, such as through differential effects on germination, growth,

Weed Biology and Climate Change, Lewis H. Ziska and Jeffrey S. Dukes, © 2011 Blackwell Publishing Ltd.

or seed production. In recent decades, many research programs have quantified crop loss by weeds through these competitive interactions. Consequently, there is an excess of data that demonstrate that a given weed species at a given density for a given crop over a given time will reduce the yield of that crop by a given amount (e.g., see Chapter 5; Zimdahl, 2004). There are of course many variations of this theme that have been explored by weed scientists, including how climate and edaphic conditions alter the degree of crop loss; the stage at which competition occurs, the spatial and temporal distribution of weeds, the influence of crop row spacing, etc.

Interestingly, it is the biological basis for yield reduction that continues to be elusive. If a weed reduces crop yield, is it because of greater extraction of a given soil nutrient relative to the crop; or, greater shading ability? Perhaps it is an indirect effect (e.g., interference) related to allelopathy, or some insect or pathogen which lives on the weed? How much does each disparate resource affect competitive outcomes? These are difficult questions to address, particularly for a field system. From a methodological point of view, how do you even separate above-ground from below-ground competition in situ? Nevertheless, one of the overarching goals of weed science is to try and distinguish general competitive traits as a means to predict weed–crop outcomes and quantify likely crop losses.

The data that are available in this regard reflect a reductionist approach to isolating and understanding an abiotic resource that is likely to affect competition. Therefore, while studies have examined the role of water, light, and temperature on crop–weed competition, few have attempted to integrate how temporal and spatial variability of these resources alter competitive outcomes. As Zimdahl (2004) emphasizes, unlike experimentalists, plants cannot separate individual resource changes during competition and any attempt to do so does not entirely reflect in situ conditions.

Still, if we are to place climate change and/or CO_2 in competitive context, it is worth briefly summarizing what is currently known regarding single environmental variables and weed–crop competition. Temperature can differentially affect photosynthesis and growth of C_3 and C_4 plants (Pearcy and Ehleringer, 1984; Sage and Kubien, 2007; Sage and Sharkey, 1987); even for plants using the same photosynthetic pathway, it can have species-specific effects on germination, early emergence, distribution, time to flowering, and senescence, with subsequent effects on competitive outcomes. Overall, stress imposed on weed–crop associations by unfavorably high or low temperatures will alter competitive outcomes (Flint and Patterson, 1983; Wall, 1993, inter alia). Weeds can reduce water availability, potentially contributing to crop stress. A survey of weed/crop competition studies in which soil water availability varied suggested a tendency for decreased water availability to favor the crop by reducing weed competition (Patterson, 1995). No data are available per se on the impact of flooding on weed/crop competition; although flooding has been practiced for centuries as a means to control weeds in rice culture. Light is recognized as the principal resource competed for by crops and weeds (Donald, 1963), and a number of critical reviews have examined the differential response of weeds and crops to light (Holt, 1995; McLachlan et al., 1993; Patterson, 1982). Although early emergence and leaf area development is important, effective competition for light is not always a question of increased leaf area, but also of plant height and shading (e.g., Joenje and Kropff, 1987) within the crop canopy. Hence, changes in light spectral quality, in addition to light quantity, may be important in determining plant height and competitive outcomes.

Weed Competition and Crop Losses with Climate Change

Carbon Dioxide

There are two available field studies that have quantified changes in seed yield with weedy competition as a function of rising CO_2 (Ziska, 2000, 2003). In these trials, two crop species, one C_3 (soybean, 30 cm rows) and other C_4 (dwarf sorghum, 40 cm rows) were grown with lambsquarters (C_3) and redroot pigweed (C_4) and velvetleaf (C_3) and redroot pigweed, respectively, at a density of two weeds per meter of row at either ambient or at ambient +250 ppm CO_2. Both weed species for either crop are recognized as being troublesome to production in a number of U.S. states (Bridges, 1992). In soybean, the loss in seed yield from lambsquarters increased (relative to a weed-free condition) from 28 to 39%; conversely, for pigweed, soybean yield losses were reduced from 45 to 30% with rising CO_2 (Figure 5.1). Velvetleaf did not affect seed yield of dwarf sorghum at ambient CO_2, but redroot pigweed reduced dwarf sorghum biomass (by ~15%). At the higher CO_2 concentration, both weedy species reduced seed yield. In both studies, increasing CO_2 resulted in taller weeds relative to the crop, suggesting that light availability may have been altered; however, the effects of above- versus below-ground competition were not examined separately.

Although these are the only studies that have examined crop yield losses from weed competition as a function of elevated CO_2 directly, several other studies have examined how rising CO_2 affects the growth of weeds versus crops (Table 5.1). Studies using C_3 crops and C_4 weeds have demonstrated that rising atmospheric CO_2 increases the ratio of crop to weed vegetative biomass, consistent with the known carboxylation kinetics of the C_3 and C_4 pathways. Consequently, rising CO_2 per se could lead to a competitive advantage for C_3 crops over C_4 weeds. However, for all other studies, the weed is favored, even when the weed and crop have the same photosynthetic pathway (Table 5.1).

Unfortunately, only a few studies have examined how CO_2 affects yield losses from weedy competition. In particular, little is known about weeds that are similar genetically or morphologically to the crop itself. Farming by its nature is designed to select the soil, temperature, moisture, and climate variables that maximize the productivity of the crop species. The attributes that make a weed more or less competitive with a crop still need to be elucidated fully, but those weeds that share physiological, morphological, or phenological traits with the crop are likely to be selected for in any consistent set of agronomic practices. Often, of course, these are weeds that are wild relatives of the domesticated crop species, for example, oat and wild oat, sorghum and shattercane, potato, and nightshade.

Do wild and cultivated biotypes of the same species respond differently to changes in CO_2, and if so, what are the subsequent consequences with respect to competition and crop yield?

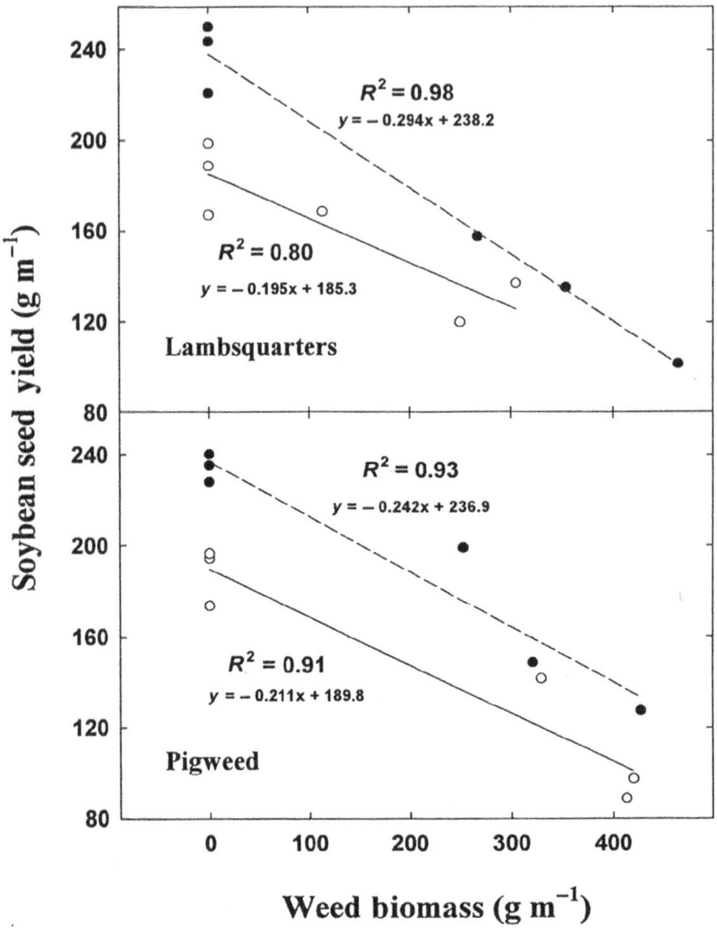

Figure 5.1 Soybean seed yield (g m^{-1} of row) as a function of increasing weed biomass (g m^{-1} of row) when grown at either ambient (open circles, solid lines) or elevated (ambient + 250 ppm, closed circles, dashed lines) carbon dioxide. No differences in the slope of the regression were observed from the study (ANCOVA; Ziska, 2000).

On the basis of the different responses of cultivated and weedy (red) rice to elevated CO_2, Ziska and McClung (2008) evaluated the competitive response using the cultivated variety "Clearfield 161" and a weed: red rice biotype "Stuttgart S." Competition was determined using a relative yield (RY) approach, which is based on the performance of two species (or biotypes) in monoculture versus mixture. In its simplest form, the RY of species A in competition with species B is the yield (or other growth parameter) per plant of A grown in mixture with B (AB) divided by the dry weight when grown in monoculture (A from AA). The change in competition (RY) can then be related to a variable change (e.g., CO_2). Elevated CO_2 reduced the RY of the cultivated variety when it was grown with red rice (Figure 5.2).

Table 5.1 Summary of studies examining whether weed or crops grown in competition were "favored" as a function of elevated CO_2.

Crop	Weed	Increasing CO_2 favors?	Environment	Reference
A. *C_4 crops/C_4 weeds*				
Sorghum	*Amaranthus retroflexus*	Weed	Field	Ziska (2003)
B. *C_4 crops/C_3 weeds*				
Sorghum	*Xanthium strumarium*	Weed	Glasshouse	Ziska (2001)
Sorghum	*Albutilon theophrasti*	Weed	Field	Ziska (2003)
C. *C_3 crops/C_3 weeds*				
Soybean	*Chenopodium album*	Weed	Field	Ziska (2000)
D. *C_3 crops/C_4 weeds*				
Fescue	*Sorghum halapense*	Crop	Glasshouse	Carter and Peterson (1983)
Soybean	*Sorghum halapense*	Crop	Chamber	Patterson et al. (1984)
Rice	*Echinochloa glabrescens*	Crop	Glasshouse	Alberto et al. (1996)
Soybean	*A. retroflexus*	Crop	Field	Ziska (2000)

"Favored" indicates whether elevated CO_2 produced significantly more crop or weed biomass.

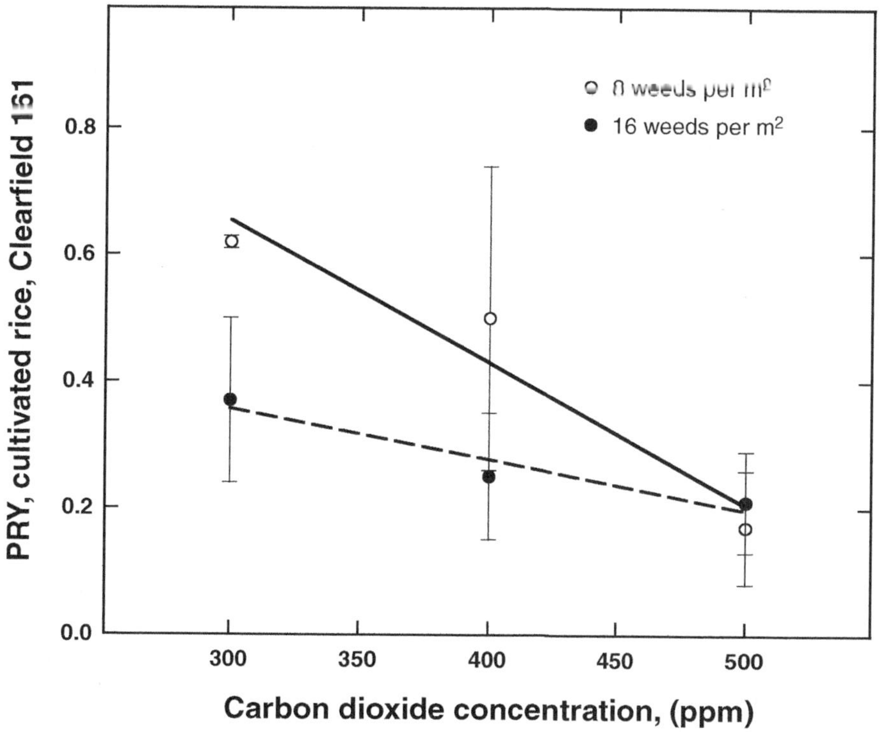

Figure 5.2 Change in plant relative yield (PRY) in cultivated rice (Clearfield 161), with different densities of red (weedy) rice (Stuttgart S) as a function of three different CO_2 concentrations. The levels used here approximate those at the beginning of the twentieth century, the end of the twentieth century, and those projected for the mid twenty-first century (Ziska, unpublished data).

Carbon Dioxide, Environmental Interactions, and Competition

Patterson et al. (1988) compared the relative effects of elevated CO_2 (700 ppm) and different day/night temperatures on growth of cotton and two C_3 weeds, velvetleaf and spurred anoda. Cotton grew faster with elevated CO_2, and this effect strengthened with increasing temperature. Only one study has evaluated the interaction between temperature, CO_2, and crop/weed competition (Alberto et al., 1996) with respect to seed yield. In this experiment, competition between rice (*Oryza sativa*, C_3) and a weedy competitor, smooth barnyardgrass (*Echinocholoa glabrescens*, C_4) was assessed at two different CO_2 (ambient and ambient +200 ppm) and two different temperatures (day/night of 27/21 and 37/29°C). At 27/21°C, increased CO_2 favored the crop (i.e., the C_3 species); however, at the higher temperature, CO_2 favored the C_4 weed, primarily because warmer conditions resulted in increased seed yield loss for rice relative to the C_4 weed (Alberto et al., 1996). Hypothetically, there are a number of additional potential interactive effects related to temperature, CO_2, and weed/crop competition. Weeds of the tropics, for example, respond strongly to small air temperature changes (Flint and Patterson, 1983; Flint et al., 1984), but it is unknown if rising CO_2 would benefit these weeds more than tropical crops in a warmer climate. The Alberto study again emphasizes that predicting how climate change will affect weed–crop competition based simply on photosynthetic pathways may be overly simplistic.

No studies are yet available on the interaction between drought, rising CO_2, and weed/crop competition. Crops and weeds grow quickly and have similar responses to drought; the overall impact of weeds may be reduced because of decreased growth of both crops and weeds in response to water availability (Patterson, 1995). An experiment in pasture found that the proportion of weed biomass increased with CO_2 to a similar extent in wet and dry treatments (although competition was not determined directly, Newton et al., 1996). If CO_2 decreases the water requirement of C_4 weeds relative to C_3 crops, C_4 weeds could still potentially out-compete C_3 crops in a high CO_2/drought condition (Knapp et al., 1993).

In settings where nutrients strongly limit plant growth, CO_2 may only minimally affect growth rates; however, under moderate or minimal nutrient limitations found in most agricultural situations, plants may respond more strongly (e.g., Rogers et al., 1993; Seneweera et al., 1994; Wong, 1979). In some species even moderate phosphate deficiencies could eliminate biomass stimulation by elevated CO_2 (e.g., Goudriaan and de Ruiter, 1983). If weeds have a greater nutrient use efficiency that the crop, then they might respond more strongly to CO_2.

In the only study examining nutrient and carbon dioxide weed:crop competition to date, Zhu et al. (2008) using a C_3 crop (rice) and a C_4 weed (barnyardgrass) model system, examined N supply and CO_2 interactions under paddy conditions. In this study, the proportion of rice biomass increased relative to barnyardgrass with a 200-ppm increase in atmospheric CO_2, but *only* if nitrogen was adequate. If N was low, elevated CO_2 reduced the competitive abilities of rice relative to the C_4 weed, presumably by reducing carbon sinks (e.g., tiller formation) in rice. These data suggest that, for rice cropping systems in which N is limited, rising CO_2 could exacerbate competitive losses, even from C_4 weeds. Schenk et al. (1997) also suggested that elevated CO_2 could alter competitive outcomes between ryegrass (Lolium spp.) and white clover depending on the supply of nitrogen. Overall, while nutrient availability is likely to modify the response of individual weeds and crops to CO_2, the relative effect on weed/crop competition is likely to be species-specific and will depend, in part, on soil–plant interactions. These include the extent of rhizobial symbiosis and N fixation, mycorrhizal colonization, nitrogen deposition, etc.

Climate Change and Competition

Aside from any direct CO_2 fertilization effect, climatic change, particularly spatial and temporal variation in precipitation and temperature are likely to have significant effects on weed–crop and weed–weed competition per se. Differential responses to temperature were observed between soybean and two weed species, sicklepod and prickly sida, in response to rising temperature, with subsequent increases in weed competition possibly resulting from differences in mycorrhizal colonization between weeds and crops (Tungate et al., 2007). Wheat is a better competitor with jointed goatgrass at low, but not high day/night temperatures (18/10 and 27/10°C, respectively; Fleming et al., 1988). Flint and Patterson (1983) report a greater enhancement of redroot pigweed growth relative to soybean with warming (26/17, 29/20, and 32/23°C day/night temperatures, respectively). Other studies demonstrating greater competitive advantage to weeds over crops include lambsquarters over sugarbeet (Houghton and Thomas, 1996) and sicklepod and Palmer amaranth over soybean (Wright et al., 1999).

With respect to weed–weed interactions, wild oat (C_3) was more competitive than green foxtail (C_4) at low relative to high temperatures (22/16 and 28/22°C, respectively) (Wall, 1993). For Queensland, Australia, temperature was a significant factor in determining competitive dominance between golden bristle grass and goosegrass (Hawton, 1979, 1980) with temperatures greater than 23°C resulting in goosegrass dominance. Even for weeds of a similar genus, differential responses of germination and growth can be observed. For example, warming accelerated germination and height growth of Powell amaranth more than that of redroot pigweed and smooth pigweed (Weaver, 1984). Responses to temperature, particularly those of plant and leaf development, will strongly affect crop–weed–weed competition and crop losses. However, projected aspects of temperature change associated with anthropogenic forcing, in particular warmer night-time temperatures, or a greater occurrence of extreme temperature events, have not been examined in regards to crop–weed competition. Plausibly, greater increases in night-time relative to day-time temperatures projected with global warming (McCarthy et al., 2001) could decrease seed production to a greater extent in crop relative to weed species (e.g., cowpea; Ahmed et al., 1993), with subsequent competition favoring the weed. Because data are lacking at present, generalizations regarding such projections on weed biology and weed–crop competition remain speculative.

A number of climatic models project not only greater desertification, but also, in some regions, increased precipitation and an increased frequency of extreme precipitation events (IPCC, 2007). How will such altered precipitation patterns alter the success of weeds, crop–weed competition, and crop losses from weeds in the future? Water is certainly an essential resource, differentially utilized by crops and weeds with a widely acknowledged role in crop–weed competition (reviewed in Zimdahl, 2004). Common cocklebur reduced yields of well-watered soybean more than of drought-stressed soybean (Mortensen and Coble, 1989). Similarly, Canada thistle had a greater impact on wheat yield with increasing rainfall (Donald and Khan, 1992). Conversely, Patterson and Highsmith (1989) demonstrated that drought increased competition between cotton, velvetleaf, or spurred anoda. Overall, response to drought in agronomic conditions is dependent on species and cultural conditions. Patterson (1995) has suggested that decreased water availability may favor the crop by reducing the competitive impact of the weed (see Table 5.1; Patterson, 1995). That is, when potential yield is already limited by water, weed competition for other resources has less impact. Water availability may also affect the duration of weed-free periods during crop development. Coble et al. (1981) demonstrated that with common ragweed, a critical period to avoid competitive effects and crop yield loss was 2 weeks in a dry year and 4 weeks in a wet year following planting. However, the duration of the critical period varied by weed and crop (Harrison et al., 1985; Jackson et al., 1985).

Unmanaged Forage Systems

These systems provide feed for livestock, and the degree of management can vary significantly, from unmanaged or lightly managed rangeland for cattle, that is, mostly unimproved land where a large proportion of the natural vegetation is natural grasses, forbs and shrubs; to pasture, which is land specifically managed to provide a high proportion of grass and alfalfa. It is important to recognize that in such systems, the concept of "weed" is fundamentally changed, in part because the characterization of "unwanted" or "undesired" is different than in highly managed crop systems, as is the degree of economic loss.

However, there are a number of weed-induced impacts on forage systems, some of which are likely to be affected by climatic change. For example, climate change is likely to impact rangelands in the Great Plains by altering the extent and variability of temperature and precipitation, with subsequent effects on productivity and quality (Derner et al., 2005). With respect to weed biology, such anticipated changes, in addition to rising atmospheric CO_2 are likely to alter species composition and the utility of rangeland for grazing (Polley et al., 1997). Shrub encroachment into many native grasslands/rangelands has been well documented and tentatively attributed (in part) to the greater sensitivity to CO_2 of C_3 shrubs than C_4 grasses (e.g., Polley et al., 1997). Recent experimental evidence supports this idea. A doubling of CO_2 shifted the plant community toward greater dominance of shrubs over a 5-year period within a short-grass steppe (Morgan et al. 2007). As such, CO_2-induced encroachment of shrubs and the subsequent reduction in grass species needed for domestic livestock can be viewed in the context of weed (i.e., shrub) populations altering the agronomic value of unmanaged land in the western U.S. (Figure 5.3).

Fire plays a dual role as natural disturbance factor and valuable tool for rangeland management. Fire can facilitate nutrient cycling and promote the growth of grasses and forbs; periodic fire maintains a number of major grassland, shrub steppe, and savanna systems.

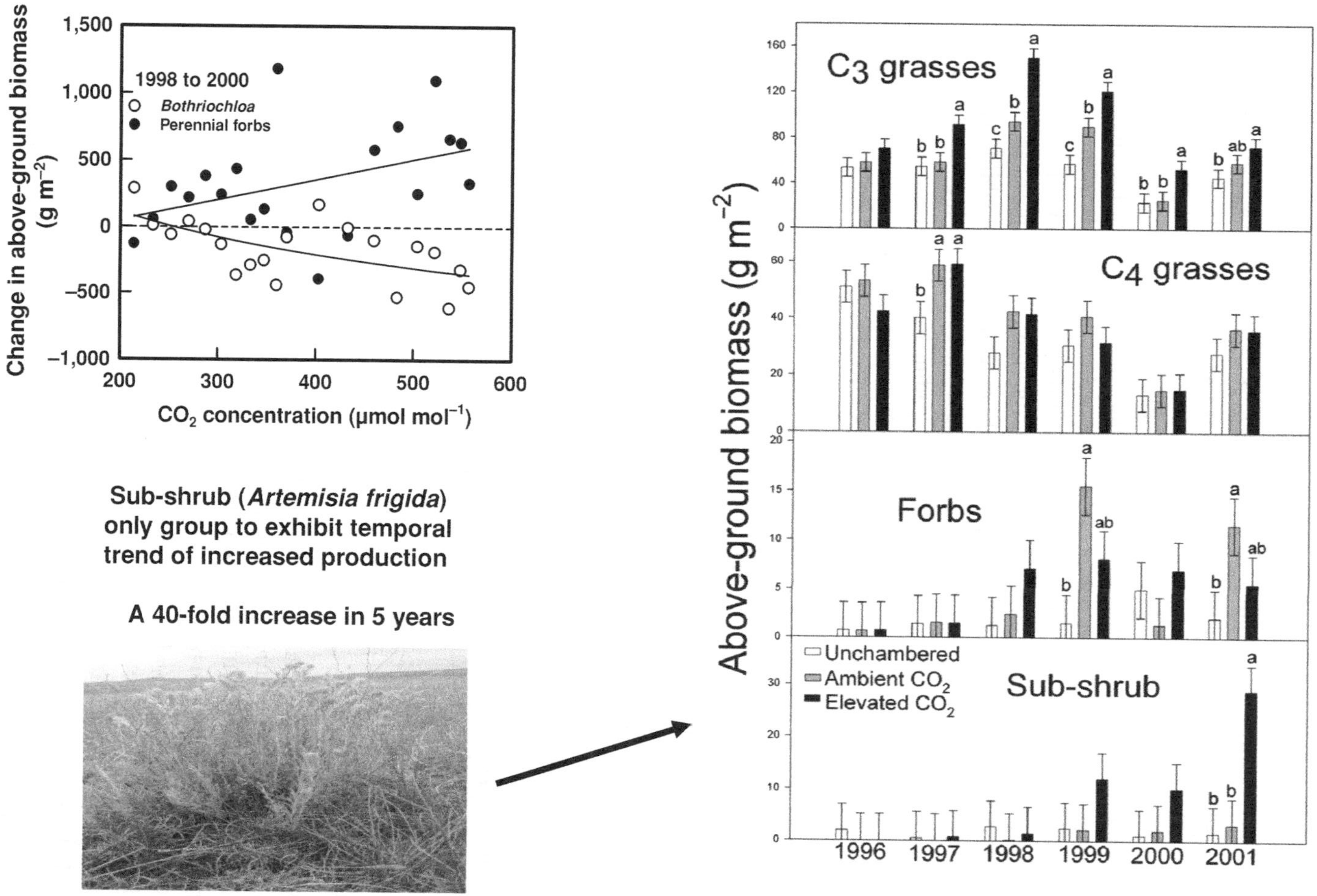

Figure 5.3 CO_2-induced changes in species diversity of rangelands with introduction of C_3 weedy shrubs.

However, when climate and vegetation increase fire frequency it can significantly reduce the growth of grasses and forbs, with significant effects on rangeland productivity.

In much of the western United States, cheatgrass has colonized open spaces between perennial, native shrubs. This central Asian annual weed grows quickly during the short wet season, leaving a carpet of flammable litter that increases the frequency of fire events during the summer (Whisenant, 1990). As fires become more frequent, nonadapted native species decline, while cheatgrass becomes increasingly dominant. Remote sensing suggests cheatgrass now dominates 20,000 km^2 (>7,700 square miles), or 7% of the Great Basin region of the western United States (Bradley and Mustard, 2005). A study of three cheatgrass populations collected from different elevations within the Sierra Nevada range revealed that even small, recent changes in atmospheric carbon dioxide (ca. 50 ppm) can increase the growth rate and combustibility of cheatgrass, while reducing its digestibility (Figure 5.4) (Ziska et al., 2005a). All of these factors can increase the amount of fine fuel on the landscape, potentially accelerating fire frequencies, and leading to further cheatgrass dominance. Cheatgrass invasions decrease the economic and environmental value of these rangelands, and lead to increased fire fighting costs (Knapp, 1996). Recent evaluations, however, of both rising CO_2 and temperature, indicate a more complex scenario, with some weeds increasing in distribution and others, such as cheatgrass, moving northward (Bradley and Mustard, 2005).

Unlike crop production systems, weed competition in pasture has not been extensively studied. Overall, dependable information on weed biology and/or cost–benefit analysis is less available to support management decisions in pastures (Penn State, 2001). It is recognized that weeds can compete with grasses or pasture to reduce both quantity and quality of available forage. For those experiments that have examined the impact of rising carbon dioxide on weeds in pasture, a similar pattern was observed as with crop systems; that is, that rising CO_2 favored the weed if the photosynthetic pathway was the same, whereas the pasture species was favored if the weed was a C_4 grass (Table 5.2). However, in general, because so little is known regarding the impact of weed species, density, duration of competition, etc. in pasture systems, it is difficult to quantify potential losses in these systems with climate or carbon dioxide.

Aquaculture

Among farming systems, aquaculture (or aquafarming), is a relatively new practice, although sporadic examples can be found back to ancient Chinese and Roman cultures (Pillay and Kutty, 2005). Overall, aquaculture, the domestication of wild aquatic species (i.e., crustacean, fish, or seaweed), has grown rapidly both in the United States and globally. Aquaculture was expected to contribute half of the total global fisheries production for human consumption in 2010 (FAO, 2008; Naylor et al., 2009). The presence of limited amounts of unwanted plant species in these systems can be of use in maintaining water quality and for food; however, unrestricted growth of aquatic weeds can affect productivity by limiting light penetration, increasing oxygen depletion or by utilization of nutrients needed for the desired species.

Although the impact of aquatic weeds is global in scope (e.g., water hyacinth is among the worst noxious weeds in Africa), only a handful of studies have examined their specific

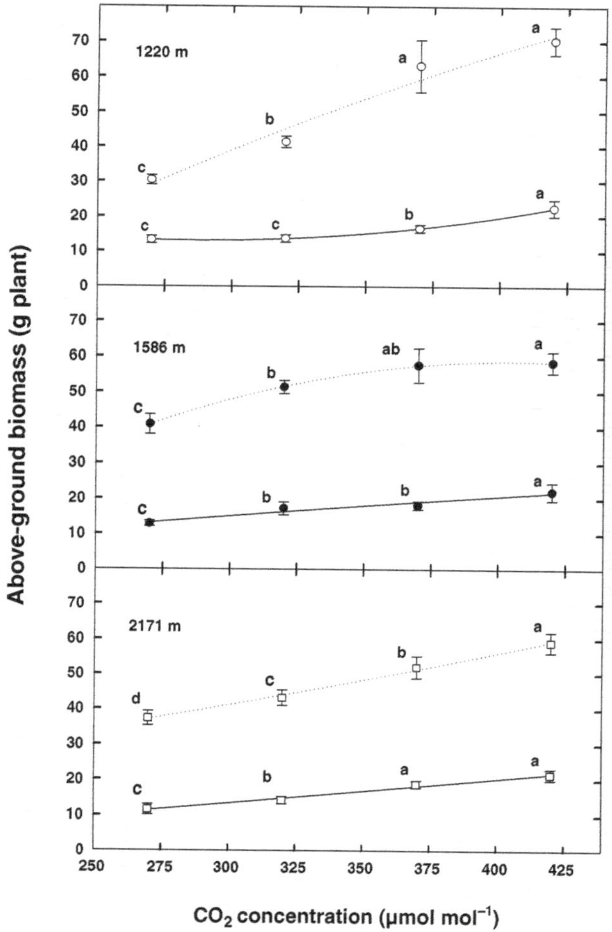

Figure 5.4 Change in above-ground biomass of three cheatgrass populations as a function of recent and near-term carbon dioxide concentrations. Bars are ±1 standard error. Picture on right shows establishment of cheatgrass monoculture in rangelands following fire events (USDA Forest Service).

Table 5.2 Summary of studies examining whether weed or pasture were "favored" as a function of elevated CO_2.

Crop	Weed	Increasing CO_2 favors?	Environment	Reference
A. *C_3 pasture/C_3 weeds*				
Lucerne	*Taraxacum officinale*	Weed	Field	Bunce (1995)
Pasture	*Taraxacum* and *Plantago*	Weed	Field	Potvin and Vasseur (1997)
Pasture	*Plantago lanceolate*	Weed	Chamber	Newton et al. (1996)
B. *C_3 crops/C_4 weeds*				
Pasture	*Paspalum dilatatum*	Crop	Chamber	Newton et al. (1996)
Lucerne	Various grasses	Crop	Field	Bunce (1993)

"Favored" indicates whether elevated CO_2 produced significantly more crop or weed biomass. "Pasture" refers to a mix of C_3 grass species.

response to climate/CO_2 and the consequences with respect to production in aquaculture. For water hyacinth, elevated CO_2 resulted in larger ramets and increased flower production (Spencer and Bowes, 1986); for hydrilla, elevated CO_2 resulted in a significant increase in plant biomass (Chen et al., 1994). Changes in water temperature, acidity, depth, fertility, and the incidence and degree of coastal flooding may also affect weeds and the productivity of aquaculture. Curiously, this area of research, particularly given the increase and interest in aquaculture, significantly lags behind our understanding of climate change and weeds in crop production systems. Clearly, there are methodological difficulties associated with experimentally manipulating CO_2 and other environmental variables for submersible vegetative plant parts, but this does not explain why no data on CO_2 and/or climate responsiveness exists for above-ground leaves in aquatic systems or for riparian weeds such as salt cedar or purple loosestrife.

Weeds on "Steroids"

The data currently available with respect to weed–crop competition and production losses suggest a greater response of the weed relative to the crop (e.g., Tables 5.1 and 5.2). Even with C_3 crop and C_4 weed comparisons, concurrent increases in both CO_2 and another parameter (e.g., temperature and nitrogen) favor the weed. So, can we surmise that weeds could do better than crops in agronomic situations with rising CO_2 and/or climate change?

To answer this question, it is important to remember that in theory, managed agriculture is implementation of an economically desirable plant monoculture. There are strong fiscal pressures to maintain homogenization in such a system including ease of harvest, uniform seed quality, pesticide application, longevity of seed storage, etc. Hence, there is an increasing trend with larger farms for greater genotypic and phenotypic uniformity within a given crop system (Doyle, 1985; Roberts, 2008). Such uniformity also allows crop selections to respond strongly to high abiotic inputs, particularly water and fertilizer in order to obtain high levels of productivity. Overall, such a strategy has been essential to maintain an adequate food supply to a global population approaching 7 billion.

But in reality, a managed agronomic system consists not only of the desired plant, but also, between 8 and 10 wild, or weedy species for a given crop species (Bridges, 1992). Both weeds and crops then are subject to selection pressures. Such pressures can be related to human selection (plant breeding), as well as environmental or natural selection. Clearly, climate change and rising CO_2 levels represent strong selection pressure within any plant community. Given the greater species (and genetic) diversity among the weed populations relative to a single crop that is genetically uniform, climate and/or CO_2 selection pressure should, hypothetically, favor weed over crop populations as per the suggestion of Treharne (1989). Unfortunately, this may be exacerbated if breeders ignore such shifting abiotic externalities (e.g., temperature and CO_2) in their own selection programs (Newton and Edwards, 2007; Ziska and McClung, 2008).

Although there is some support for this hypothesis from individual weed–crop competition studies, it is unclear if such competitive pressures are evident for assemblages of different plant species either in managed or unmanaged systems. Multi-year assessments of how climate change and/or rising CO_2 alter plant competition are rare. There are a handful of studies that have addressed the selection question with respect to projected carbon dioxide increases. A comparison of the impact of increasing CO_2 concentration on yellow starthistle demonstrated a significant increase in biomass in monoculture, but a nonsignificant impact when yellow starthistle was grown within a nutrient-poor serpentine grassland community (Dukes, 2002), suggesting that, in this context, rising carbon dioxide did not stimulate the growth of this plant species preferentially. However, in another more fertile type of California grassland, yellow starthistle responded much more strongly than competing species to elevated CO_2, particularly in disturbed microsites (Dukes et al. unpublished). Work with the invasive species honey mesquite and the native species little bluestem suggests that the woody invader, honey mesquite, responds more strongly to CO_2 (Polley et al., 1994, 2002). Research on Japanese honeysuckle in a forest under-story also demonstrated a strong CO_2 growth response and subsequent increase in percent cover (Belote et al., 2003) in one out of 2 years, with similar trends in the second year. In that same forest, the invasive grass *Microstegium vimeneum* did nor respond to CO_2. A weedy tree species in Switzerland showed a stronger CO_2 response than native trees (Hattenschwiler and Korner, 2003), also suggesting preferential growth of a nonnative species. Finally, elevated CO_2 increased the productivity and success of a rangeland weed associated with fire outbreaks in an arid ecosystem (Smith et al., 2000). Overall, four of the five seminal studies suggest that rising levels of CO_2 can preferentially benefit "weed" species. If this holds true in other systems, crop losses could be exacerbated as atmospheric carbon dioxide increases.

Pests and Disease: Gimme Shelter

Carbon Dioxide, Climate Change, and Pathogens

Weeds in cropping systems can serve as reservoirs for plant pathogens. As such, weeds can act as the initial inoculum for crop epidemics, or can exacerbate existing problems by providing multiple hosts for reproduction. For example, black nightshade is a host for *Phytophthora infestans* (responsible for the Irish potato famine of the 1840s). Some nightshade species respond strongly to CO_2 (Flynn et al., 2006). Kudzu, a widely established weed in the southeastern United States, also responds strongly to rising CO_2 levels (Sasek and Strain, 1988). Kudzu, in turn, serves as an alternate host for Asian soybean rust, a pathogen that threatens soybean production in the southeastern and Midwestern states. In general, CO_2-induced stimulation of weed biomass could increase the mass of host tissue for infection and use by pathogens; conversely, larger plants might tolerate more severe levels of infection without subsequent reductions in seed production.

For agroecosystems, climate and/or CO_2 could increase density and canopy height with a subsequent increase in humidity and potential changes in the fecundity of leaf-infecting fungi (e.g., rusts and powdery mildew; Chakraborty and Datta, 2003). Increased crop canopy residues at the end of the growing season could also influence pathogenic overwintering (Manning and Tiedemann, 1995) while CO_2-induced changes in the ratio of carbon to nitrogen (C:N) could influence the degree of fungal sporulation during decomposition of that material. Malmstrom and Field (1997) report that CO_2 enrichment resulted in increased persistence of wild oat plants infected with barley yellow dwarf virus (BYDV) in chamber experiments. Overall, several responses to rising carbon dioxide could alter disease susceptibility and weed–crop interactions, but there are no field-based observations of the effects of rising CO_2 on pathogen biology (Chakraborty and Datta, 2003; Sutherst et al., 2007).

In addition to effects of carbon dioxide on pathogen fecundity, climate change could alter the ability of weeds to act as carriers for disease. Disease development overall is favored by warm, moist conditions. Winter is a significant factor in pathogen mortality, with more than 99% of pathogen populations experiencing mortality (Burdon and Elmqvist, 1996). Rising winter temperature may remove this thermal limitation for pathogen morbidity with a subsequent increase in their range and distribution. Mild winters and warmer weather have been associated with increased outbreaks of powdery mildew, leaf spot disease, leaf rust, and rizomania disease, all of whom have weed hosts. Warmer temperatures as already discussed are likely to change weed distribution as well, with subsequent effects on pathogen ranges and potential introduction of new disease in cooler regions (Treharne, 1989). However, the overall effect is complex. For example, increased precipitation per se could increase disease spread because wet leaf surfaces are necessary for spore germination and infection to proceed. Conversely, climatic extremes (e.g., floods and drought) could lessen disease problems by reducing the establishment and fecundity of the weed host.

To complicate matters, it is also likely that concurrent increases in CO_2 and climate are likely to interact to alter micrometeorological parameters in ways that will be difficult to predict. For example, rising CO_2 can increase water use efficiency and may moderate drought stress by closing stomata. Differential responses between weeds and crops may, in turn, alter inoculation and spread of diseases that enter through stomata, or that require a given relative water content within the leaf. Alternatively, drought avoidance could help to maintain secondary plant chemicals that could be used to ward off pathogens (Coakley et al., 1999).

Carbon Dioxide, Climate, and Insects

As with weeds and pathogens, insects and their weed hosts are likely to respond to carbon dioxide and climate in ways that are likely to have positive and negative impacts on crop production. Weeds are a recognized host for many insect pests (Capinera, 2005). Because some insects feed exclusively on weeds (e.g., Colorado potato beetle on hairy nightshade; Horton and Capinera, 1990), changes in climate and/or carbon dioxide that result in increased insect feeding on weed hosts could benefit crop production by reducing weed competition. Alternatively, there are a number of insects who are not restricted to a single weed host, and can move from weeds to crops with a subsequent increase in crop damage and reduced yield. The degree of taxonomic similarity between weeds and crops may be a factor in predicting crop damage by weed-feeding insects (Capinera, 2002).

Carbon dioxide-induced changes in leaf quality are also likely to also have positive and negative effects on insect behavior and mortality (Bezemer and Jones, 1998; Lincoln et al., 1993). For example, elevated CO_2 led to increased growth and development of *Polyommatus icarus* (Lepidoptera) larvae feeding on bird's foot trefoil, a common weed in North America (Goverde et al., 1999), due in part to increases in leaf digestibility and carbohydrate. However, buckeye butterfly grew more slowly and had higher mortality when they fed on a lawn weed (common plantain) that was grown in elevated CO_2. Plantain leaves were less nutritious, with lower N concentrations (Osbrink et al., 1987). Leaf-sucking insects could also experience CO_2-associated changes in leaf quality. At least some mites benefit from CO_2 possibly as a consequence of increased nonstructural carbohydrates in the leaf (Heagle et al., 2002). Population density of aphids increased on groundsel and annual blue grass (Bezemer and Jones, 1998).

The basis for shifts in insect fecundity with carbon dioxide is not entirely clear. Higher C:N ratios associated with increasing CO_2 could result in compensatory increases in foliar consumption rates by insects as a means to obtain nitrogen. Increased consumption rates in turn could be accompanied by a decrease in the efficiency of plant tissue conversion to body mass and reduction in insect growth rate. Conversely, it is possible that higher C:N ratios could have a negative impact on plant secondary compounds (allelochemicals), that act as a deterrent for insect feeding (e.g., Kerslake et al., 1998). For example, elevated levels of carbon dioxide inhibited plant defenses in soybean when attacked by an insect, Japanese beetle (*Popillia japonica*; Zavala et al., 2008). Many weeds are hosts to adult Japanese beetles and grassy or weedy areas surrounding fields may act as potential breeding grounds for the pest. At the community level, the overall impact of increasing carbon dioxide is difficult to predict, and will likely depend on long-term changes in competition and plant community structure (weed species shifts). Overall, most data have focused on a single insect–plant species interaction, making a more complex assessment of weed/crop interactions and crop damage difficult.

Given that temperature is among the central regulating factors controlling the development and fecundity of agronomic insects, climate warming is likely to increase their geographical distribution (both in latitude and altitude) and/or enhance their capacity for over-wintering. While long-term records are rare, aphid flight activity has been continuously recorded in the United Kingdom since 1964. These data indicate accelerated flight phenology by 3–6 days, corresponding to an increase of 0.4°C in mean temperature from 1964 (Fleming and Tatchell, 1995). Similar analyses across a number of aphid species project an advance in the timing of their spring migration by 2 weeks for every 1°C increase in winter temperature (Harrington, 2002). While temperature is likely to affect insect fecundity and hence crop damage directly; range expansion of insects will be dependent on host expansion, and weed and crop hosts are likely to expand their ranges in the future. If temperature expands the range of insect pest, this could bring them into contact with other potential crop or weed hosts present at distant locations. This has been suggested by Cannon (1998) who examined the spread of invasive insects in northwestern Europe and the implications for insect invasions in the United Kingdom. On the basis of the fossil record, a sharp increase in insect herbivory occurred during previous geological warming periods (Currano et al., 2008).

Changes in precipitation frequency and amount are also likely to alter insect/host relationships in agricultural systems (Fuhrer, 2003). For example, intense precipitation can serve as a deterrent to oviposition by insects (e.g., European corn-borer; Davidson and Lyon, 1987). Intense precipitation and flooding may also disadvantage insect species that spend part of their life cycle underground (Watt and Leather, 1986).

Alternatively, drought could make weeds or crops more attractive to insects by concentrating carbohydrates (Mattson and Haack, 1987). Shifting precipitation could also influence the occurrence and population density of weed and crop hosts (e.g., Drake, 1994). As with pathogens, direct physiological effects of rising carbon dioxide in conjunction with climatic extremes will almost certainly alter the biology of insect pests and subsequent interactions between agricultural weeds and crop hosts, but in ways that will be difficult to currently project. In addition, there is little information on how weeds, insects, and pathogens are likely to interact in the future, even though it is recognized that insects can carry pathogens.

Summary

As emphasized in Chapter 1, plants are classified as weeds in part because of the harm that they inflict on human systems. In this chapter, we have reviewed how rising carbon dioxide and climate change could modify the degree of harm induced by weeds at the agricultural or farm level. Given that weeds remain a major constraint to agricultural production, there is evidence from a number of studies that competition and crop limitations could increase from rising carbon dioxide and climate variability in precipitation and temperature. We have proposed that the greater genetic variability and number of weeds relative to a single genetically uniform crop monoculture may reflect the ability of weeds to successfully adapt to rapid changes in the environment. We have also examined the indirect harm associated with weeds as hosts for insects and pathogens and have surmised a number of potential mechanisms as to how carbon dioxide and/or climate variability could alter the ability of weeds or crops to act as carriers for these pests. While a number of specifics are lacking, particularly at the ecological level, it is clear that weed impacts in agricultural communities can be expected to change. In the next two chapters, we will continue the theme of how climate and carbon dioxide can modify weed damage, examining how climate change is likely to alter invasions of natural ecosystems, and weeds and public health.

References

Ahmed FE, Hall AE, Madore MA. 1993. Interactive effects of high temperature and elevated carbon dioxide concentration on cowpea [*Vigna unguiculata* (L.) Walp.]. *Plant Cell Environ.* **16**:835–842.

Alberto AMP, Ziska LH, Cervancia CR, Manalo PR. 1996. The influence of increasing carbon dioxide and temperature on competitive interactions between a C_3 crop, rice (*Oryza sativa*) and a C_4 weed (*Echinochloa glabrescens*). *Aust.* J. Plant Physiol. **23**:795–802.

Belote RT, Weltzin JF, Norby RJ. 2003. Response of an understory plant community to elevated CO_2 depends on differential responses of dominant invasive species and is mediated by soil water availability. *New Phytol.* **161**:827–835.

Bezemer TM, Jones TH. 1998. Plant–insect herbivore interactions in elevated atmospheric CO_2: quantitative analyses and guild effects. *Oikos* **82**:212–222.

Bradley BA, Mustard JF. 2005. Identifying land cover variability distinct from land cover change: cheatgrass in the Great Basin. *Remote Sens. Environ.* **94**:204–213.

Bridges DC. 1992. *Crop Losses Due to Weeds in the United States.* Weed Science Society of America, Champaign, IL, 403 pp.

Bunce JA. 1993. Growth, survival competition and canopy carbon dioxide and water vapor exchange of first year alfalfa at an elevated CO_2 concentration. *Photosynthetica* **29**:557–565.

Bunce JA. 1995. Long-term growth of alfalfa and orchard grass plots at elevated carbon dioxide. *J. Biogeo.* **22**:341–348.

Burdon JJ, Elmqvist T. 1996. Selective sieves in the epidemiology of *Melampsora lini*. *Plant Pathol.* **45**:933–943.

Cannon RJC. 1998. The implications of predicted climate change for insect pests in the UK, with emphasis on non-indigenous species. *Global Change Biol.* **4**:785–796.

Capinera JL. 2002. North American vegetable pests: the pattern of invasion. *Am. Entomol.* **48**:20–39.

Capinera JL. 2005. Relationships between insect pests and weeds: an evolutionary perspective. *Weed Sci.* **53**:892–901.

Carter R, Peterson KM. 1983. Effects of a CO_2-enriched atmosphere on the growth and competitive interaction of a C_3 and a C_4 grass. *Oecology* **58**:50–54.

Chakraborty S, Datta S. 2003. How will plant pathogens adapt to host plant resistance at elevated CO_2 under changing climate? *New Pathol.* **159**:733–742.

Chen D-X, Coughenour MB, Eberts D. Thullen J. 1994. Interactive effects of CO_2 enrichment and temperature on the growth of dioecious *Hydrilla verticillata*. *Environ. Exp. Bot.* **34**:345–353.

Clements FE, Weaver JE, Hanson HC. 1929. *Plant Competition: An Analysis of Community Functions.* Carnegie Institution of Washington [Publication 398], Washington, DC.

Coakley SM, Scherm H, Chakraborty S. 1999. Climate change and plant disease management. *Ann. Rev. Phytopath.* **37**:399–426.

Coble HD, Williams FM, Ritter RL. 1981. Common ragweed (*Ambrosia artemisiifolia*) interference in soybeans (Glycine max). *Weed Sci.* **29**:339–342.

Currano ED, Wilf P, Wing SL, Labandeira CC, Lovelock EC, Royer DL. 2008. Sharply increased insect herbivory during the Paleocene-Ecocene thermal maximum. *Proc. Natl. Acad. Sci.* **105**:1960–1964.

Davidson RH, Lyon WF. 1987. *Insect Pests of Farm, Garden and Orchard.* John Wiley & Sons, New York, 640 pp.

Derner JD, Schuman GE, Jawson MD, Shafer SR, Morgan JA, Polley HW, Runion GB, Prior SA, Torbert HA III, Rogers HH Jr., Bunce JA, Ziska LH, White JW, Franzluebbers AJ, Reeder SJ, Venterea RT,

Harper LA. 2005. USDA-ARS global change research on rangelands and pasturelands. *Rangelands* **27**:36–42.

Donald CM. 1963. *Competition Among Crop and Pasture Plants. Advances in Agronomy*, Vol. 15. Academic Press, New York, pp. 1–118.

Donald WW, Khan M. 1992. Yield loss assessment for spring wheat (*Triticum aestivum*) infested with Canada thistle (*Cirsium arvense*). *Weed Sci.* **40**:590–598.

Doyle J. 1985. Altered harvest. *Agriculture, Genetics, and the Fate of the World's Food Supply*. Viking Press, New York, 502 pp.

Drake VA. 1994. The influence of weather and climate on agriculturally important insects—an Australian perspective. *Aust. J. Agric. Res.* **45**:487–509.

Dukes JS. 2002. Comparison of the effect of elevated CO_2 on an invasive species (*Centaurea solstitialis*) in monoculture and community settings. *Plant Ecol.* **160**:225–234.

Fleming GF, Young FL, Ogg AG Jr. 1988. Competitive relationships among winter wheat (*Triticum aestivum*), jointed goatgrass (*Aegilops cylindrica*), and downy brome (*Bromus tectorum*). *Weed Sci.* **36**:479–486.

Fleming RA, Tatchell GM. 1995. Shifts in the flight periods of British aphids: a response to climate warming? In: Harrington R, Stork NE (eds) *Insects in a Changing Environment*. Academic Press, London, pp. 505–508.

Flint EP, Patterson DT. 1983. Interference and temperature effects on growth in soybean (*Glycine max*) and associated C_3 and C_4 weeds. *Weed Sci.* **31**:193–199.

Flint EP, Patterson DT, Mortensen DA, Reichers GH, Beyers JL. 1984. Temperature effects on growth and leaf production in three weed species *Weed Sci.* **32**:655–663.

Flynn DFB, Suddereth EA, Bazzaz FA. 2006. Effects of aphid herbivory on biomass and leaf level physiology of *Solanum dulcamara* under elevated temperature and CO_2. *Environ. Exp. Bot.* **56**:10–18.

Food and Agriculture Organization of the United Nations 2008. FISHSTAT Plus: Universal Software for Fishery Statistical Time Series (Food and Agriculture Organization, Rome) Version 2.32.

Fuhrer J. 2003. Agro-ecosystem responses to combinations of elevated CO_2, ozone and global climate change. *Agric. Ecosys. Environ.* **97**:1–20.

Goudriaan J, de Ruiter HE. 1983. Plant growth in response to CO_2 enrichment, at two levels of nitrogen and phosphorous supply. I. Dry matter, leaf area and development. *Neth. J. Agric. Sci.* **31**:157–169.

Goverde M, Bazin A, Shykoff JA, Erhardt A. 1999. Influence of leaf chemistry of *Lotus corniculatus* (Fabacea) on larval development of *Polyommatus icarus* (Lepidoptera, Lycaenidae): effects of elevated CO_2 and plant genotype. *Funct. Ecol.* **13**:801–810.

Harrington R. 2002. Insect pests and global environmental change. In:Douglas I (ed.) *Encyclopedia of Global Environmental Change*, Vol. **3**. Wiley, Chichester, UK, pp. 381–386.

Harrison SK, Williams CS, Wax LM. 1985. Interference and control of giant foxtail (*Setaria faberi*) in soybeans (Glycine max). *Weed Sci.* **33**:203–208.

Hattenschwiler S, Korner C. 2003. Does elevated CO_2 facilitate naturalization of the non-indigenous *Prunus laurocerasus* in Swiss temperate forests? *Funct. Ecol.* **17**:778–785.

Hawton D. 1979. Temperature effects on *Eleusine indica* and *Setaria anceps* grown in association (I). *Weed Res.* **19**:279–284.

Hawton D. 1980. Temperature effects on *Eleusine indica* and *Setaria anceps* grown in association (II). *Weed Res.* **20**:261–266.

Heagle AS, Burns JC, Fisher DE, Miller JE. 2002. Effects of carbon dioxide enrichment on leaf chemistry and reproduction by two-spotted spider mites (Acari: Tetrachynidae) on white clover. *Environ. Entomol.* **31**:594–601.

Holt JS. 1995. Plant responses to light: a potential tool for wed management. *Weed Sci.* **43**:474–482.

Horton DR, Capinera JL. 1990. Host utilization by Colorado potato beetle (Coleoptera: Chrysomelidae) in a potato/weed (*Solanum sarrachoides* Sendt.) system. *Can. Entomol.* **122**:113–121.

Houghton SK, Thomas TH. 1996. Effects of elevated carbon dioxide concentration and temperature on the growth and competition between sugarbeet (*Beta vulgaris*) and fat-hen (*Chenopodium album*). *Aspect Appl. Biol.* **45**:197–204.

IPCC: Climate Change 2007. *Impacts, Adaptation and Vulnerability*. IPCC Secretariat, Geneva, Switzerland, 986 pp.

Jackson LA, Kapusta G, Schutte-Mason DJ. 1985. Effect of duration and type of natural weed infestations on soybean yield. *Agron. J.* **77**:725–729.

Joenje W, Kropff MJ. 1987. Relative time of emergence, leaf area development and plant height as major factors in crop–weed competition. *Brit. Crop. Prot. Conf. Weeds* **3**:971–978.

Kerslake JE, Woodin SJ, Hartley SE. 1998. Effects of carbon dioxide and nitrogen enrichment on a plant–insect interaction: the quality of *Calluna vulgaris* as a host for *Operophtera brumata*. *New Phytol.* **140**:43–53.

Knapp AK, Hamerlyn CK, Owensby CE. 1993. Photosynthetic and water relations response to elevated CO_2 in the C_4 grass, *Andropogon gerardii*. *Int. J. Plant Sci.* **154**:459–466.

Knapp PA. 1996. Cheatgrass (*Bromus tectorum* L.) dominance in the Great Basin Desert: history, persistence, and influences to human activities. *Global Environ. Change* **6**:37–52.

Lincoln DE, Fajer ED, Johnson RH. 1993. Plant insect herbivore interactions in elevated CO_2 environments. *TREE* **8**:64–68.

Malmstrom CM, Field CB. 1997. Virus-induced differences in response of oat plants to elevated carbon dioxide. *Plant Cell Environ.* **20**:178–188.

Manning WJ, Tiedemann AV. 1995. Climate change: potential effects of increased atmospheric carbon dioxide (CO_2), ozone (O_3) and ultraviolet-B (UV-B) radiation on plant diseases. *Environ. Pollut.* **88**:219–245.

Mattson WJ, Haack RA. 1987. The role of drought in outbreaks of plant-eating insects. *Biosciences* **37**:110–118.

McCarthy JJ, Canziani OF, Leary NA, Dokken DJ, White KS. 2001. *Climate Change 2001: Impacts, Adaptation & Vulnerability. Contribution of Working Group II to the Third Assessment Report of the Intergovernmental Panel on Climate Change*. Cambridge University Press, Cambridge, 1000 pp.

McLachlan SM, Tollenaar M, Swanton CJ, Weise SF. 1993. Effect of corn-induced shading on dry matter accumulation, distribution and architecture of redroot pigweed (*Amaranthus retroflexus*). *Weed Sci.* **41**:568–573.

Morgan JA, Milchunas DG, LeCain DR, West M, Mosier AR. 2007. Carbon dioxide enrichment alters plant community structure and accelerates shrub growth in the shortgrass steppe. *Proc. Natl. Acad. Sci. U. S. A.* **104**:14724–14729.

Mortensen DA, Coble HD. 1989. The influence of soil water content on common cocklebur (*Xanthium strumarium*) interference in soybeans (*Glycine max*). *Weed Sci.* **37**:76–83.

Naylor RL, Hardy RW, Bureau DP, Chiu A, Elliott M, Farrell AP, Forster I, Gatlin DM, Goldburg RJ, Hua K, Nichols PD. 2009. Feeding aquaculture in an era of finite resources. *Proc. Natl. Acad. Sci. U. S. A.* **106**(36):15103–15110.

Newton PCD, Clark H, Bell CC, Glasgow EM. 1996. Interaction of soil moisture and elevated CO_2 on the above-ground growth rate, root length density and gas exchange of turves from temperate pasture. *J. Exp. Bot.* **47**:771–779.

Newton PCD, Edwards GR. 2007. Plant breeding for a changing environment. In:Newton PCD, Carran RA, Edwards GR, Niklaus PA (eds) *Agroecosystems in a Changing Climate*. CRC Press, Boca Raton, FL, pp. 309–319.

Oerke EC. 2006. Crop losses to pests. *J. Agric. Sci.* **144**:31–43.

Osbrink WLA, Trumble JT, Wagner RE. 1987. Host suitability of *Phaseolus lunata* for *Trichoplusia ni* (Lepidoptera: Noctuidae) in controlled carbon dioxide atmospheres. *Environ. Entomol.* **16**:639–644.

Patterson DT. 1982. Shading responses of purple and yellow nutsedges (*Cyperus rotundus* and *C. esculentus*). *Weed Sci.* **30**:25–30.

Patterson DT. 1995. Effects of environmental stress on weed/crop interactions. *Weed Sci.* **28**:71–75.
Patterson DT, Highsmith MT. 1989. Competition of spurred anoda (*Anoda cristata*) and velvet leaf (*Albutilon theophrasti*) with cotton (Gossypium hirsutum) during simulated drought and recovery. *Weed Sci.* **37**:658–664.
Patterson DT, Flint EP, Beyers JL. 1984. Effects of CO_2 enrichment on competition between a C_4 weed and a C_3 crop. *Weed Sci.* **32**:101–105.
Patterson DT, Highsmith MT, Flint EP. 1988. Effects of temperature and CO_2 concentration on the growth of cotton (*Gossypium hirsutum*), spurred anoda (*Anoda cristata*), and velvetleaf (*Abutilon theophrasti*). *Weed Sci.* **36**:751–756.
Pearcy RW, Ehleringer JR. 1984. Comparative ecophysiology of C_3 and C_4 plants. *Plant Cell Environ.* **7**:1–13.
Pennsylvania State Extension Bulletin 2001. Agronomy Facts No. 52; Weed Management in Pasture Systems, 15 pp.
Pillay TVR, Kutty MN. 2005. *Aquaculture: Principles and Practices*, 2nd edn. Blackwell Publishing, Oxford, UK, 624 pp.
Polley HW, Johnson HB, Mayeux HS. 1994. Increasing CO_2: comparative responses of the C_4 grass *Schizachyrium* and grassland invader *Prosopis*. *Ecology* **75**:976–988.
Polley HW, Mayeux HS, Johnson HV, Tischler CR. 1997. Viewpoint: atmospheric CO_2, soil water, and shrub/grass ratios on rangelands. *J. Range Manage.* **50**:278–284.
Polley HW, Johnson HB, Tischler CR. 2002. Woody invasion of grasslands: evidence that CO2 enrichment indirectly promotes establishment of *Prosopis glandulosa*. *Plant Ecol.* **164**:85–94.
Potvin C, Vasseur L. 1997. Long-term CO_2 enrichment of a pasture community. *Ecology* **78**:666–677.
Roberts P. 2008. *The End of Food*. Houghton Mifflin Press, New York, 390 pp.
Rogers GS, Payne L, Milham P, Conroy J. 1993. Nitrogen and phosphorus requirements of cotton and wheat under changing atmospheric CO_2 concentrations. *Plant Soil* **156**:231–234.
Sage RF, Kubien DK. 2007. The temperature response of C_3 and C_4 photosynthesis. *Plant Cell Environ.* **30**:1086–1107.
Sage RF, Sharkey TD. 1987. The effect of temperature on the occurrence of O_2 and CO_2-insensitivity photosynthesis in field grown plants. *Plant Physiol.* **84**:658–664.
Sasek TW, Strain BR. 1988. Effects of carbon dioxide enrichment on the growth and morphology of Kudzu (*Pueraria lobata*). *Weed Sci.* **36**:28–36.
Schenk U, Jager HJ, Weigel HJ. 1997. The response of perennial ryegrass/white clover swards to elevated atmospheric CO_2 concentrations. 1. Effects on competition and species composition and interaction with N supply. *New Phytol.* **135**:67–79.
Seneweera S, Milham P, Conroy J. 1994. Influence of elevated CO_2 and phosphorus nutrition on the growth and yield of a short-duration rice (*Oryza sativa* L. cv jarrah). *Aust. J. Plant Physiol.* **21**:281–292.
Smith SD, Huxman TE, Zitzer SF, Charlet TN, Housman DC, Coleman JS, Fenstermaker LK, Seemann JR, Nowak RS. 2000. Elevated CO_2 increases productivity and invasive species success in an arid ecosystem. *Nature* **408**:79-82.
Spencer W, Bowes G. 1986. Photosynthesis and growth of water hyacinth under CO_2 enrichment. *Plant Physiol.* **82**:528–533.
Sutherst RW, Baker RHA, Coakley SM, Harrington R, Kriticos DJ, Scherm H. 2007. Pests under global change—meeting your future landlords? In:Canadell JC, Pataki DE, Pitelka LF (eds) *Terrestrial Ecosystems in a Changing World*. Springer, Berlin Heidelberg, pp. 211–226.
Treharne K. 1989. The implications of the 'greenhouse effect' for fertilizers and agrochemicals. In:Bennet RC (ed.) *The Greenhouse Effect and UK Agriculture*. Ministry of Agriculture, Fisheries and Food, UK, pp. 67–78.
Tungate KD, Israel DW, Watson DM, Rufty TW. 2007. Potential changes in weed competitiveness in an agroecological system with elevated temperatures. *Environ. Exp. Bot.* **60**:42–49.

Wall DA. 1993. Comparison of green foxtail (*Setaria viridens*) and wild oat (*Avena fatua*) growth, development and competitiveness under three temperature regimes. *Weed Sci.* **41**:369–378.

Watt AD, Leather SR. 1986. The pine beauty in Scottish lodgepole pine plantations. In:Berryman AA (ed.) *Dynamics of Forest Insect Populations: Patterns, Causes, Implications*. Plenum Press, New York, pp. 243–266.

Weaver SE. 1984. Differential growth and competitive ability of *Amaranthus retroflexus, A. powelli* and *A. hybridus*. *Can. J. Plant Sci.* **64**:715–724.

Whisenant SG. 1990. Changing fire frequencies on Idaho's snake river plains: ecological and management implications. In:McArthur ED, Romney EM, Smith SD, Tueller PT (eds) *Proceedings of the Symposium on Cheat Grass Invasion, Shrub Die-Off, and Other Aspects of Shrub Biology and Management: 4–10.* Intermountain Research Station, United States, Forest Service, Ogden, Utah.

Wong SC. 1979. Elevated atmospheric partial pressure of CO_2 and plant growth I. Interactions of nitrogen nutrition and photosynthetic capacity in C_3 and C_4 plants. *Oecol* **44**:68–74.

Wright SR, Coble HD, Raper CD Jr., Rufty TW Jr. 1999. Comparative responses of soybean (Glycine max), sicklepod (*Senna obtusifolia*), and Palmer amaranth (*Amaranthus palmeri*) to root zone and aerial temperatures. *Weed Sci.* **47**:167–174.

Zavala JA, Casteel CL, DeLucia EH, Berenbaum MR. 2008. Anthropogenic increase in carbon dioxide compromises plant defense against invasive insects. *Proc. Natl. Acad. Sci. U. S. A.* **105**:5129–5133.

Zhu C, Zeng Q, Ziska LH, Zhu J, Xie Z, Liu G. 2008. Effect of nitrogen supply on carbon dioxide induced changes in competition between rice and barnyardgrass (*Echinochloa crus-galli*). *Weed Sci.* **56**:66–71.

Zimdahl RL. 2004. *Weed Crop Competition: A Review*. Blackwell Publishing, Oxford, UK, pp. 27–106.

Ziska LH. 2000. The impact of elevated carbon dioxide on yield loss from a C_3 and C_4 weed in field grown soybean. *Global Change Biol.* **6**:899–905.

Ziska LH. 2001. Changes in competitive ability between a C_4 crop (Sorghum bicolor L.) and a C_3 weed, common cocklebur (*Xanthium strumarium* L.) with elevated carbon dioxide. *Weed Sci.* **49**:622–627.

Ziska LH. 2003. Evaluation of yield loss in field sorghum from a C_3 and C_4 weed with increasing CO_2. *Weed Sci.* **51**:914–918.

Ziska LH, McClung A. 2008. Differential response of cultivated and weedy (red) rice to recent and projected increases in atmospheric carbon dioxide. *Agron. J.* **100**:1259–1263.

Ziska LH, Reeves JB III, Blank R. 2005a. The impact of recent increases in atmospheric CO_2 on biomass production and vegetative retention of Cheatgrass (Bromus tectorum): implications for fire disturbance. *Global Change Biol.* **11**:1325–1332.

6 Invasive Plants and Climate Change in Natural Ecosystems

"Weed 'em and Weep"

Anonymous

Introduction

As discussed in Chapter 1, "weed" is a subjective term, identifying an unwanted plant species. It is a term in common use, particularly in an agricultural context. Nonnative plant species that invade *natural* ecosystems and cause ecological or economic harm, while meeting the classic definition of a weed as a "plant that does harm," are often referred to as "invasive plant species," "nonnative invasive plant species," or one of a variety of similar terms (Colautti and MacIsaac, 2004). Such distinctions occur perhaps because ecologists find these terms more objective, or to distinguish them from the classic agricultural weeds—indeed, many invasive plant species were imported purposefully for their beauty. Since this chapter mainly concerns natural ecosystems, we use "invasive plant species" here, abbreviated IPS, rather than "invasive weeds" to conform to common ecological language.

Invasive Plants in Natural Ecosystems: Ecology and Significance

Since we have primarily considered the economic and environmental impact of agricultural weeds to this point, it is worth briefly considering some similarities and differences in the processes through which weeds invade agricultural systems and IPS invade wildlands. Invasion can be viewed as a process that occurs in stages (e.g., Theoharides and Dukes, 2007; Vermeij, 1996), beginning with the transport of a species to a new region, followed by its initial survival in and colonization of the region, establishment of persistent, spreading populations, and finally spread of new populations throughout the landscape. For IPS and agronomic weeds, the initial transport stage can occur through many different pathways, including purposeful and accidental introductions. Many IPS were introduced as ornamental species, or for soil conservation purposes (OTA, 1993). These purposefully introduced species were often brought from regions with similar climates to overcome the major abiotic barrier associated with colonization and establishment, in which individuals must tolerate local environmental conditions to survive and reproduce.

Weed Biology and Climate Change, Lewis H. Ziska and Jeffrey S. Dukes, © 2011 Blackwell Publishing Ltd.

While the first stages of invasion may share some common features in agricultural and natural systems, the recipient systems have major differences. Most obviously, natural ecosystems are typically disturbed less frequently and in different ways than agroecosystems, and are less subject to direct manipulation of the plant species growing on the landscape. Natural areas do not typically receive direct inputs of fertilizer, and are less likely to be candidates for chemical control, although chemicals can be used to control the worst invaders in many areas. Also, natural areas are typically less vigilantly observed for the arrival and spread of unwanted plant species than agricultural fields. These conditions permit a different group of invasive species to thrive than those found in agricultural settings—a group that includes both woody and herbaceous perennials.

With such major differences in selective forces between managed and unmanaged systems, different traits are associated with invasiveness. For example, traits associated with IPS include use by humans, the capacity to spread through vegetative means, early flowering and long flowering intervals, high fecundity and germination rate, high specific leaf area (typically meaning thin leaves), tolerance of herbivory or clipping, and tolerance of fire (Pyšek and Richardson, 2007). Of course, the degree to which each of these traits actually promotes invasiveness depends to some extent on the ecosystem in which a species finds itself—for instance, fire tolerance is only an asset in systems that burn.

Just as some traits contribute to invasiveness, certain characteristics of ecosystems may make them more (or less) susceptible to invasion, or at least can determine which IPS are most likely to become established. Unless an IPS can tolerate the climatic conditions and disturbance regime in a new location, it will not persist. One of the fathers of invasion biology, Charles Elton, noted that highly disturbed areas were typically more heavily populated with invasive species than undisturbed areas (Elton, 1958). In less disturbed systems, IPS may encounter communities in which dozens of other plant species already coexist at high densities and compete for available resources. Disturbances often make available resources that could allow IPS to become established, at least for a brief window of time (Davis et al., 2000). However, disturbances in natural areas can also keep some IPS out. In contrast to weed species in agricultural settings, IPS in natural systems may encounter frequent fire, shifting substrate, flooding, animal disturbance, and other factors.

Many ecologists have wondered how IPS can become wildly successful outside of their native ranges when they often do not appear particularly aggressive in their home ranges, and several researchers have developed and tested a range of hypotheses on this subject.

One straightforward hypothesis is that IPS fill an empty niche; an ecosystem may have available resources that native species, by chance of history or ancestry, have not evolved to exploit as rapidly or as fully. In this case, an IPS with the right traits may be able to exploit the resource, and in doing so outcompete or coexist with the natives. This mechanism likely contributes to the success of many IPS (e.g., Dyer and Rice, 1999; Shea and Chesson, 2002).

A second hypothesis suggests that IPS are "released" from their natural enemies when they arrive in a new habitat; the pathogens and herbivores that plague them in their native range are likely to be absent from other continents, and so they are able to grow faster. Meanwhile, native species suffer under their own load of natural enemies, reducing their competitive abilities. This advantage of IPS is implied, although not conclusively demonstrated, when an introduced biocontrol agent (a natural enemy from the native range) attacks an IPS, successfully reducing its population (Keane and Crawley, 2002). Many biocontrol species have effectively controlled IPS, but there are also many cases in which biocontrol has failed (McFadyen, 1998).

The enemy release hypothesis has led to theories that IPS can evolve, becoming better competitors as they allocate less of their resources to defense against natural enemies, and more to growth (Blossey and Notzold, 1995). These hypotheses have found support in some studies (e.g., Joshi and Vrieling, 2005), but the process does not appear to have occurred everywhere (Genton et al., 2005). Another hypothesis based on enemy release suggests that the species that have the highest enemy loads are those living in the most resource-rich environments, and so the species adapted to high resource availability are likely to benefit the most from enemy release (Blumenthal et al., 2009). This would also help to explain the greater success of IPS in high-resource environments (Daehler, 2003; Dukes and Mooney, 1999; Scherer-Lorenzen et al., 2007).

A somewhat different hypothesis, though similar in many ways to enemy release, suggests that some IPS release chemical exudates that favor IPS in competition with native plants, either through direct effects that the plants have not evolved to tolerate, or through effects on the microbial community that alter native plant growth (Callaway and Ridenour, 2004). Although this "novel weapons" hypothesis has been implicated as a factor in the success of diffuse knapweed (Callaway and Aschehoug, 2000; Vivanco et al., 2004) and spotted knapweed (Bais et al., 2003, but see Blair et al., 2009; Duke et al., 2009), how frequently this mechanism occurs in nature is unknown. Another herbaceous species, garlic mustard (Prati and Bossdorf, 2004; Stinson et al., 2006), is known to slow the growth of some of its mycorrhizal competitors by releasing compounds from its roots that inhibit arbuscular mycorrhizae.

The rate of spread of an IPS through the landscape will be a function of its establishment success, the degree to which suitable patches of habitat are connected throughout the landscape, and the dispersal ability of the IPS through that landscape, which will depend on its mode of reproduction and, for some IPS, the presence of seed-dispersing mutualists. "Disturbance corridors" such as roads, trails, and rights-of-way for power lines can facilitate rapid dispersal of IPS, because they often provide a habitat with relatively high resource availability in which many IPS have a competitive advantage, and because some corridors serve as pathways for dispersal vectors such as cars, horses, or people (D'Antonio et al., 2000).

Once they become dominant, IPS can have a variety of impacts on native species and ecosystem processes in natural systems. When native plant species get outcompeted by IPS, this often means poorer habitat quality for native animals (Dukes and Mooney, 2004). Although there are few documented extinctions of native species caused by IPS (Gurevitch and Padilla, 2004), IPS can cause radical changes in population sizes of native plant and animal species and the functioning of the systems they enter.

Invasive plants can alter virtually all important ecosystem processes, including primary productivity, decomposition, nutrient cycling, hydrology, and even geomorphological processes. These impacts have been detailed in hundreds of studies, and summarized in many review papers (e.g., Dukes and Mooney, 2004; Levine et al., 2003). A few of the more dramatic impacts include the acceleration of fire cycles by cheatgrass and other invasive grasses (Brooks et al., 2004; D'Antonio and Vitousek, 1992), reorientation and restructuring of North American sand dunes by European beachgrass (Wiedemann and Pickart, 1996), and changes in the cycling of soil nutrients in many locations (Ehrenfeld, 2003).

Just as weeds have economic impacts in agroecosystems, IPS can have economic consequences in natural ecosystems. While the major economic damage from weeds of crops and pastures comes from crop damage and loss (~\$25 billion out of \$33 billion in annual costs in the United States alone), the major cost of many IPS in natural areas comes from control efforts (\$100 million per year for IPS in U.S. aquatic systems, out of \$110 million total costs; Pimentel et al., 2005). The overall economic impact of IPS in natural systems far exceeds \$200 million annually in the United States, and may exceed that value for a single species (e.g., Melaleuca or "punk" trees; Serbesoff-King, 2003), but comprehensive estimates are not yet available (Duncan et al., 2004; Pimentel et al., 2005). Many of these economic consequences are associated with IPS control, but others are associated with the impacts of IPS on natural resources. For instance, some expect salt cedar (*Tamarix* spp.) invasions to decrease hydropower production, availability of irrigation and municipal water, and river recreation revenue in the southwestern United States, leading to projected losses of \$7–16 billion in ecosystem services over 55 years (Zavaleta, 2000).

Because IPS can disrupt natural ecosystems, many land managers work diligently to identify invasions early and eradicate the IPS on their properties. There are some basic differences in how IPS are typically addressed in natural systems compared to weed management regimes used in intensively managed agroecosystems. These differences stem from several differences between the types of sites, including levels of vigilance, accessibility of sites, acceptable side effects of control strategies, and budgets for control. Natural areas often spread across a patchwork of land owners, each with their own management plans (or lack thereof). Some landowners may ignore IPS, allowing problems to intensify, while others may be exceptionally diligent about removing them.

Management techniques for IPS are tailored to the location and the IPS. Biocontrol agents have been released for several species, but rarely achieve 100% success. Prescribed fire, targeted grazing, herbicides, and manual pulling are all common control methods for IPS in natural areas. In some cases, IPS removal areas are replanted or sown with native species following treatment. In general, government agencies and landowners rarely have the resources to completely control IPS in the natural areas they manage. There is a clear advantage to early detection and treatment in terms of minimizing costs, but early detection relies largely on communication among land managers in a region. A group of "Invasive Plant Councils" has sprung up in the United States to increase communication and facilitate IPS control at the regional level (http://www.naeppc.org/), and these have in general been quite successful.

Environmental Changes and Invasive Species

Later this century, natural ecosystems will experience markedly different conditions than they do today, and the conditions they experience today already differ from those in the recent past. Throughout the world, terrestrial plant communities are encountering warmer climates and substantially higher concentrations of CO_2. In some locations, precipitation patterns are changing. In other locations, more nitrogen or more ozone are entering plants and soil from the atmosphere (with contrasting effects)—and these are just some of the more obvious changes. Invasive species continue to be transported to new regions, and land continues to be converted from forest to agriculture to urban areas, or sometimes revert back. So, how will IPS fare as these changes continue?

Warmer and Weedier: Effects of a Changing Climate

Climate change alters three parameters that determine the prevalence of IPS: abiotic conditions, biotic conditions, and human behavior. Climate change itself directly alters the abiotic, or physical, conditions experienced by ecological communities. As resident communities respond to a new climatic regime, range shifts, and physiological adjustments by species change the biotic conditions experienced by any new entrant to the community, and as society prepares for and responds to climate change, some human behaviors change, with a variety of possible consequences for IPS.

Responses to Changing Abiotic Conditions

As climate change alters the environmental conditions under which species compete, conditions may shift away from those that favor native plant species. This has the potential to favor IPS that arrive preadapted to the new climatic regime (Byers, 2002), but could also disadvantage IPS in some cases (Figure 6.1).

For both native plants and IPS, climate change is likely to have its greatest initial influence at the edges of species' distributions. In general, warming should allow ranges to expand poleward and upward in elevation, but cause them to contract at the warmer edges. Model projections for changes in precipitation are more variable than they are for warming, and projected changes vary by region (IPCC, 2007). However, many species' ranges are constrained by soil moisture at some times of year, particularly in arid and semiarid regions. There, any change in precipitation regime could strongly affect species' ranges.

Several researchers have attempted to predict how IPS ranges may respond to climate change using climate envelope models paired with projections of future climate. Climate envelope models estimate the areas that are climatically suitable for a species based on information about its climatic tolerances. This information is often inferred from a species' current range. Areas are then identified that are likely to be suitable for that species in a future climate. Some of the models that consider IPS (as opposed to native species) have developed a set of suitable climate conditions based on the native range, with the thought that the species may still be expanding in its introduced range (e.g., Sutherst et al., 1995). However, research indicates that some IPS occupy different climatic space in native versus introduced ranges (Beaumont et al., 2009; Kriticos and Randall, 2001). This suggests that predictive models should be informed by meteorological information from occupied areas in the introduced range.

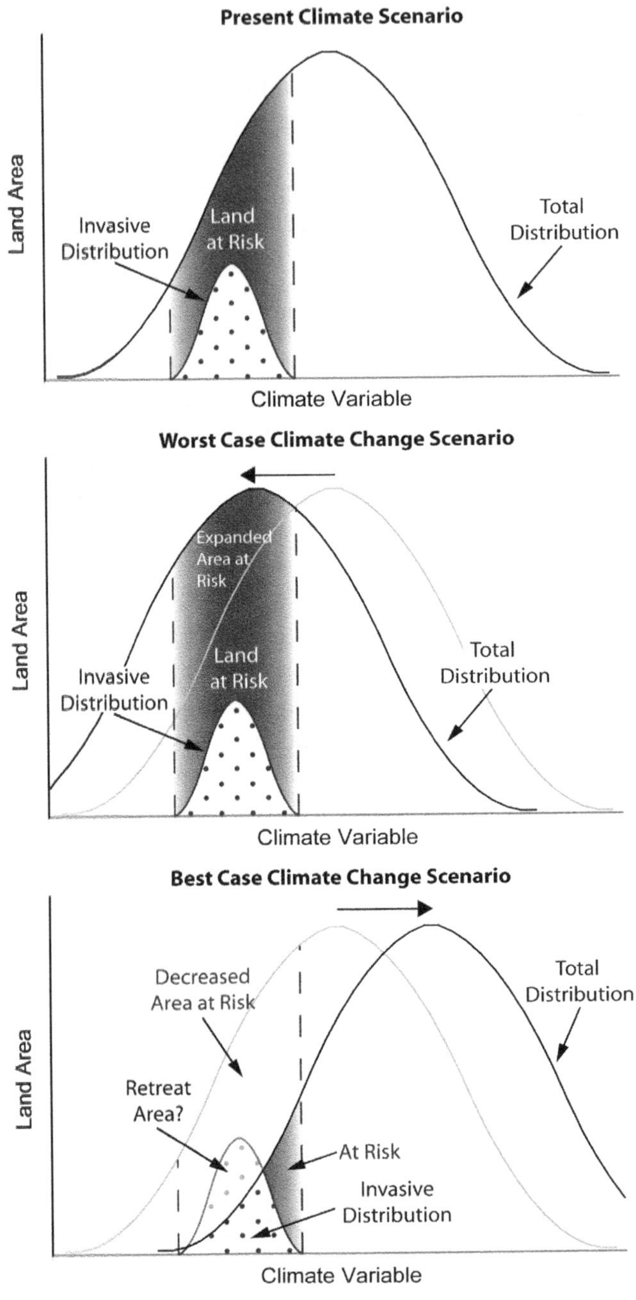

Figure 6.1 Conceptual diagram of how climate change may affect the potential range sizes of IPS. Currently, (a) distributions of IPS (shown in black dots) fill a subset of the climatically suitable land area (between dotted lines), with the remaining area (gray) at risk for future invasion. Climate change will shift more of the land area in a region toward (b) or away from (c) the conditions to which the IPS is best adapted. A shift toward the climatic optimum of the IPS would increase its potential range size. A shift away from its climatic optimum would decrease the potential range size of the IPS, and possibly even shift the IPS out of its climatically suitable range in some locations. Adapted from Bradley et al. (2009).

Climate envelope models provide a useful mechanism for making general predictions about how species' ranges may shift, but they have a variety of pitfalls. First, many plants' ranges depend not only on climate but also on soil type, which is rarely considered in such models. Second, the models cannot take into account how competitive interactions, or a dependence on mutualistic interactions, may restrict ranges. In addition, models that determine climate space based on current distributions can be problematic if they depend on the native range (as discussed above), or if the species is still expanding its range in the introduced region. Finally, these models do not account for dispersal mechanisms of the modeled species, which could influence their rates of spread, and most do not consider the direct and indirect responses of the species to other aspects of global environmental change. Nonetheless, the results of some of these studies, offer the best predictions available of where invasive species could potentially exist in the future, and where they will likely become less abundant.

Research to date suggests that potential ranges of IPS will respond to climate change in a variety of ways, from large increases to large decreases. In the United States, the work of Bradley et al. (2009) suggests that some IPS will shift ranges, while other IPS ranges will expand or contract. The IPS expected to increase range size were yellow starthistle (Figure 6.2) and salt cedar (*Tamarix* spp.), while leafy spurge was expected to contract and ranges of cheatgrass and spotted knapweed were expected to shift with little change in net range area. A model used by Jarnevich and Stohlgren (2009) suggested that kudzu may increase its range in the northeastern United States while becoming less abundant in some parts of the Southeast. Bradley and colleagues (2009) suggest that when models identify areas likely to become unsuitable for IPS, those areas could be targeted for preemptive restoration, before other unwanted species replace the currently dominant IPS.

Outside the United States, researchers have developed a similarly mixed outlook for IPS responses. Simulations by Beerling et al. (1995) suggest that the range of Japanese knotweed will shift northward in Europe. The work of Richardson et al. (2000b) that takes soil-type restrictions into account suggests that range sizes will decrease for five of South Africa's worst IPS. Kriticos and colleagues (2003a, 2003b) foresee large increases in the potential distributions of two IPS in Australia. All of these predictions depend on the accuracy of climate projections and have many of the caveats discussed above. However, the general patterns identified in these modeling exercises can be used to identify likely future challenges and opportunities.

Abiotic shifts in resource availability such as increases in atmospheric nitrogen deposition and changes in water availability may also alter the conditions in which plant species compete. Many studies have shown that IPS compete well in resource-rich conditions (Daehler, 2003; Scherer-Lorenzen et al., 2007). Davis et al. (2000) captured this concept in their theory of fluctuating resource availability (Figure 6.3). This hypothesis suggests that, where precipitation becomes more abundant, increased water availability could increase the success of IPS. However, climate models suggest that future precipitation will come in fewer, larger events, which could increase plant water stress, particularly in mesic regions (Knapp et al., 2008).

Additional studies have examined the influence of water availability on the success of IPS. By applying supplemental water to plots of Colorado shortgrass steppe for 5 years, Milchunas and Lauenroth (1995) fostered the establishment of populations of nonnative

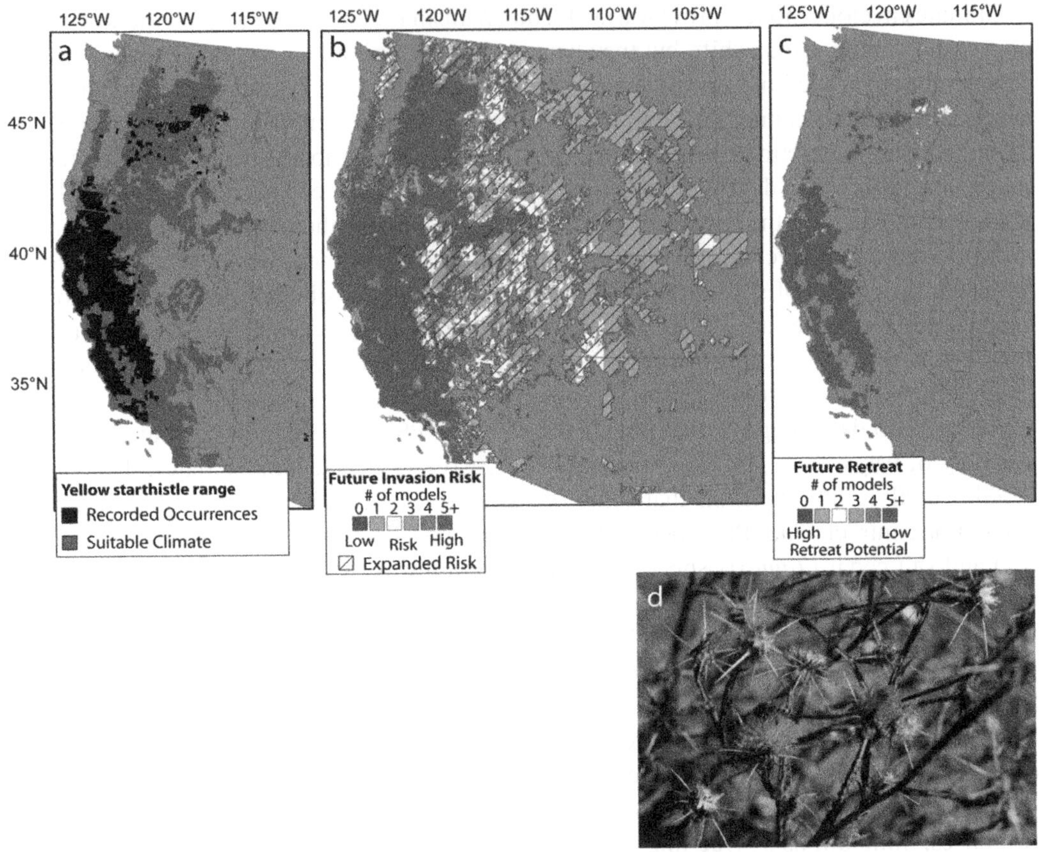

Figure 6.2 Current and projected ranges of yellow starthistle (*Centaurea solstitialis*; [a–c]) and the plant in bloom (d). Panels (a–c) from Bradley et al. (2009). (a) Current range of yellow starthistle (black) and climatically suitable habitat (red) in the western United States at the end of the twentieth century. (b) Number of climate models (out of ten models used) predicting habitat suitability by the end of the twenty-first century. Black lines show areas where invasion risk increases. (c) Areas of yellow starthistle's current range in which climatic suitability decreases (lighter grey) or stays the same (darker grey). Model simulations suggest most of yellow starthistle's current range will remain suitable. Panels (a–c) available in color online at: http://people.umass.edu/bethanyb/research.html.

plants that remained in their plots 16 years later. Blumenthal and colleagues (2008) added snow to mixed grass prairie plots in Wyoming and found that three of six IPS they had seeded in benefited, while native species did not.

Responses to Changing Biotic Conditions

Independent of any specific climatic regime, a shifting climate by itself may favor species with certain traits, and IPS may benefit as a result. Several studies suggest that IPS have wider climatic tolerances than noninvasive species (Figure 6.4; Goodwin et al., 1999; Qian and Ricklefs, 2006; Rejmánek, 1995). IPS could benefit from this tolerance by becoming more

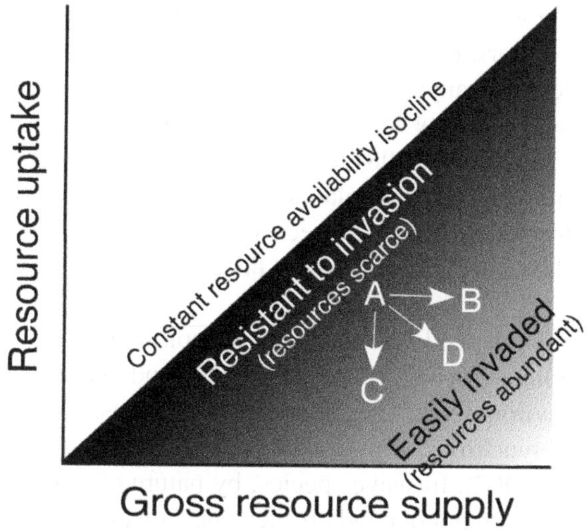

Figure 6.3 Conceptual diagram of factors affecting resource availability and plant community invasibility. As the rate of soil resource supply exceeds that of resource uptake by a plant community, that community is thought to become susceptible to invasion. An increase in resource availability, through a pulse of water or nitrogen for example, could move the system from point A to B. A disturbance that kills resident plants, such as overgrazing, could move a system from A to C. Disturbances that also create a pulse of resources, such as a fire that kills resident plants and leaves a nutrient-rich ash, could move the system from A to D. Any of these transitions would make the system more easily invaded. Adapted from Davis et al. (2000).

dominant as climate change disproportionately stresses their competitors. Such tolerance has already been shown to benefit nonnative arthropods (Chown et al., 2007).

Other traits are likely to benefit IPS under a changing climate, as well. Potential ranges for many species may shift faster than native species can track them (Malcolm et al., 2002). Consequently, species that can rapidly colonize newly habitable ranges via long-distance seed dispersal and short juvenile periods, should be at an advantage. Many IPS can disperse

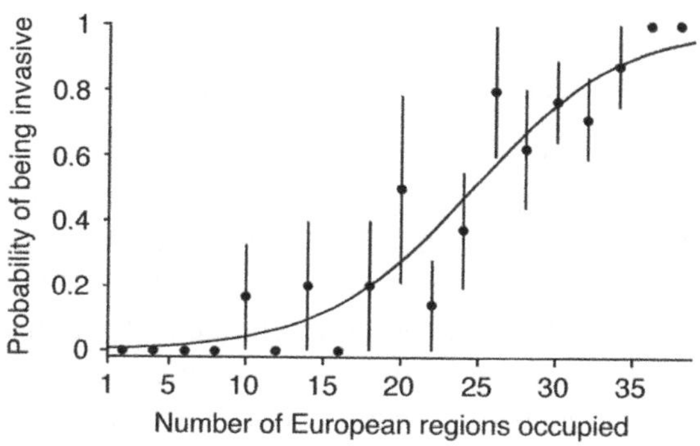

Figure 6.4 Probability (mean ± SE) of a European plant species being a successful invader of New Brunswick, Canada, based on the number of geographic ranges the species occupies in Europe. Species occupying more ranges (and thus tolerating a wider range of environmental conditions) were more likely to be invasive. Adapted from Goodwin et al. (1999).

over long distances, either via light, wind-dispersed seeds (Rejmánek and Richardson, 1996) or with the aid of animals, cars, or other vectors (Lonsdale and Lane, 1994).

Those species that do shift ranges rapidly will likely have a further advantage over the natives upon their arrival in a new location: less limitation by other trophic levels. In a greenhouse study, Engelkes et al. (2008) found that, among riparian species in the Netherlands, recently arrived, range-expanding plants experienced less herbivory aboveground and less negative soil feedbacks than native species. If this pattern of localized "enemy release" is widespread, then it could facilitate a faster replacement of native vegetation as climates shift.

Plant species that do not depend on specialists for pollination or seed dispersal may also have an advantage as climates change. Plants that depend on specialist mutualisms may respond differently to environmental cues than their mutualists, so a changing climate could bring the species out of synchrony with each other's phenological stages, to the detriment of both (Memmott et al., 2007). Invasive species, by nature of having flourished outside of their native range, typically do not depend on such specialist interactions (Richardson et al., 2000a).

Responses to Changing Human Behaviors

People will respond to climate change in a variety of ways, sometimes with consequences for IPS. For instance, planting of biofuel crops could lead to introduction of invasive species in some regions, if these crops are not carefully chosen with regard to potential environmental impacts (Barney and DiTomaso, 2008; Raghu et al., 2006). In addition, people may take advantage of new transport pathways, for instance if the Northwest Passage across the Arctic Ocean becomes navigable by cargo ships. This could affect IPS if some of these species depend on short transport times to remain viable upon arrival in a new location, or if this changes the magnitude of trade (and exchange of seeds and plants) between different ports.

Next Stop, Weed Paradise?

Implications of a Changing Atmosphere for Two Problem Species

As discussed in previous chapters, many invasive species respond positively to increasing atmospheric CO_2 concentrations, particularly when they are grown in isolation. But IPS rarely grow in isolation, and CO_2 concentration affects more than just plant growth. Many factors will interact to determine whether CO_2 and other variables favor IPS in the future (Patterson and Flint, 1990). For example, leaves exposed to elevated CO_2 transpire more slowly than those in today's atmosphere, leaving wetter soils in some ecosystems. In many systems, plants grow faster under elevated CO_2, which could affect the risk of fire in seasonally dry systems. Also, herbivores and other animals can respond to CO_2 by changing their interactions with plants. These indirect consequences of CO_2 could play a role in determining which species ultimately benefit from a changing atmosphere. We now examine some of these complexities in two case studies of problematic plant species that are both agricultural weeds and IPS in natural systems: Canada thistle and cheatgrass.

Canada thistle can be found among crops and within natural areas throughout much of North America. Skinner and colleagues (2000) found that state and provincial noxious weed lists in the United States and Canada mentioned Canada thistle more often than any other species. We have discussed many other agricultural weed species that respond to CO_2 in previous chapters. Unlike most of these, Canada thistle is a perennial species, capable of regenerating from very small root fragments (Prentiss, 1889). Perhaps as a result of this perennial habit, the weed responds more strongly to CO_2 than many other species in growth chamber studies (Ziska, 2003a). Ziska (2003a) found the species was more responsive to the recent change from preindustrial to modern conditions (+180% response of plant biomass) than to the anticipated change from modern to late twenty-first century conditions (+72%), suggesting this species may currently be so pernicious in part because of the change in atmospheric composition.

For many plant species, tissue nitrogen (N) concentrations decrease when they are grown under elevated CO_2 (Cotrufo et al., 1998). In some cases, nitrogen availability may limit the species' growth responses to CO_2 (Poorter et al., 1996). In a growth chamber experiment, Ziska (2003b) examined growth responses of Canada thistle to preindustrial versus current CO_2 concentrations when grown as individual plants at different levels of N availability, and found that the relative response of various growth parameters to CO_2 was not affected by N levels in the potting medium.

As we mentioned in earlier chapters, cheatgrass (also known as downy brome) is a widespread IPS that has invaded the Great Basin desert and floristically similar areas to its north, as well as wastelands in many other parts of North America. Its growth form is nothing like that of Canada thistle; it is an annual grass, and is generally short-statured and short-lived. Also different: cheatgrass causes more problems in death than in life. The persistent litter from cheatgrass can form a continuous fine fuel layer in the Great Basin patches that it invades, and this fuel can carry fire farther than it previously would have reached. This property of the species has dramatically accelerated the fire cycle in the Great Basin (Stewart and Hull, 1949; Whisenant, 1990), leading to conversion of areas dominated by sagebrush and other fire-intolerant shrubs to areas dominated by cheatgrass. Increased cheatgrass dominance further contributes to the problem, with economic consequences via increased firefighting costs (Knapp, 1996), and in some cases reduced forage value (Rice, 2005).

Recently, analyses of satellite-acquired data suggested that, by the mid-1990s, cheatgrass had become dominant on 20,000 km^2 of the Great Basin, comprising 7% of the land cover in the region (Bradley and Mustard, 2005). There are a couple of scenarios in which cheatgrass could become even more dominant in western North America. One would be through changes in climate that favor its growth, at least in occasional years. Cheatgrass tends to grow well in wet years (Stewart and Hull, 1949), building up a litter layer that keeps fire risks elevated for subsequent years. Thus, we would expect wetter winters to favor the species in many locations. By estimating climatic tolerances of cheatgrass and comparing them to climates projected by a range of climate models for later this century, Bradley (2009) proposed some areas of possible range growth for the species, and some areas of likely range retreat. But these predictions do not incorporate the responses of cheatgrass and other species to the changing atmosphere.

As discussed in the previous chapter, cheatgrass grows much faster when exposed to elevated CO_2, under controlled conditions (Smith et al., 1987; Ziska et al., 2005). This suggests that atmospheric changes over the last 150 years may have contributed to the ability of this species to increase fire frequencies in the western United States; more plant growth would mean greater fuel load accumulation and consequent increases in the amount of time invaded areas were susceptible to fire, the rate of fire spread in those areas, and the intensity of fires.

Supplemental CO_2 also reduces the digestibility and protein content of cheatgrass, accelerates flowering in some populations, and increases root:shoot ratios (Ziska et al., 2005). These changes could increase the proportion of biomass left standing at the end of the growing season, either by reducing herbivory or accelerating regrowth. However, these results come from growth chambers, examining plants grown in much more resource-rich conditions than typically are found in the field. A related species, red brome, responded strongly to elevated CO_2 in field conditions in the southern Nevada desert, but only in a wet El Niño year (Smith et al., 2000). In subsequent dry years, germination of the species was very poor. It is possible that cheatgrass would, similarly, respond to elevated CO_2 only in occasional years where other resources did not limit growth. On the other hand, fires may provide subsequent resource-rich conditions in which CO_2 would be more likely to stimulate growth.

In a recent greenhouse study, three IPS from China responded more strongly to CO_2 than their indigenous co-occurring species or congeners (Song et al., 2009). As in Ziska (2003a), these plants were grown as individual species in pots. If these results can be extrapolated to competitive situations (a somewhat uncertain exercise; Poorter and Navas, 2003), then this study similarly suggests that rising CO_2 by itself may contribute to IPS success.

Fertile Future for IPS

Increases in anthropogenic N deposition are likely to accompany climate change in many regions. The sources of incoming N vary by region, but primary sources are related to fossil fuel use and agriculture. Fossil fuel combustion leads to release of reactive nitrogen contained in the fossil fuels, and also creates new reactive N from atmospheric N_2 during the combustion process. Fertilized agricultural fields also contribute to the load of reactive N in the atmosphere, primarily through emission of ammonia and NO_x (Galloway et al., 2008). Regions downwind (or downstream) of urban, industrial, or agricultural areas receive the greatest influx of new reactive N. Cumulatively, human inputs of reactive N to terrestrial ecosystems exceed N inputs from natural processes, and continue to increase (Galloway et al., 2008; Gruber and Galloway, 2008).

Fertilization with N affects the composition and functioning of natural systems, and has been found to increase invasibility of some ecosystems, particularly those in which native species are adapted to resource-poor conditions (Dukes and Mooney, 1999; Scherer-

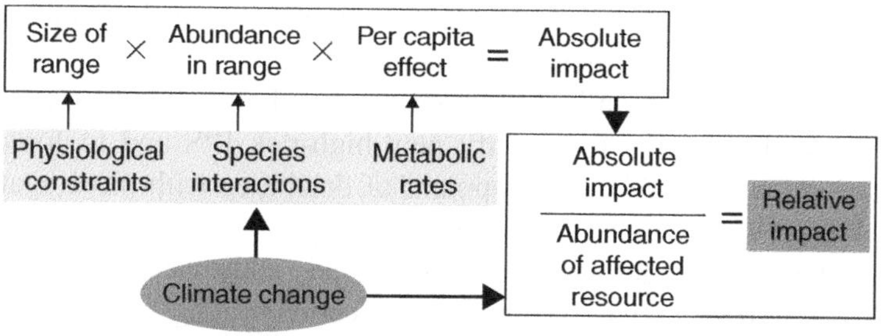

Figure 6.5 Mechanisms through which a changing climate can affect the impacts of invasive species. Reprinted from Dukes (2011).

Lorenzen et al., 2007). North American examples of these include serpentine grasslands in California (Huenneke et al., 1990), where N deposition may already have contributed to an increased abundance of invasive annual grasses (Weiss, 2000), Minnesota grassland (Wedin and Tilman, 1996), and Mojave desert, where nitrogen deposition favored introduced annual grasses over native species in a wet year (Brooks, 2003).

Future Impacts of IPS

As environmental conditions change, so will the impacts of IPS. A popular equation for estimating the overall impact of an IPS on a given variable suggests that the cumulative impact of a species can be calculated by multiplying its range, average density within that range, and its per-capita effect on the variable of interest (e.g., Parker et al., 1999). Under climate change, each of these variables could be altered. In addition, the abundance or frequency of the variable itself could change, affecting the relative impact of the IPS (Figure 6.5). An instructive example comes from the southwestern United States, where shrubby salt cedar trees line many rivers and streams. Although there is dispute over the trees' impacts (Shafroth et al., 2005), some studies have suggested that the invasion of salt cedar into the region decreased river flows, as the salt cedar caused more water to be lost from the system through transpiration (e.g., Weeks et al., 1987). Even if environmental changes do not affect the range, density, or per-capita effect of salt cedar on water resources, the predicted decrease in precipitation for the region (Karl et al., 2009) will make water more scarce, increasing its value, and increasing the relative impact of the IPS (Zavaleta, 2000).

Summary

Land managers cope with uncertainty on several levels, and climate change has increased the level of abiotic and biotic uncertainty (Lawler et al., 2010). Changes in the atmosphere and climate bring new challenges for land managers, some of which are associated with IPS. These challenges include: (1) the arrival of new IPS, (2) the increasing invasiveness of some nonnative species, (3) the decreasing invasiveness of some IPS, and possibly (4) the decreasing competitiveness of resident native communities.

To address the first of these challenges, managers can build on existing monitoring programs, databases, and networks to identify new high-risk IPS and to forecast changing invasion fronts (Lee et al., 2008). Development of databases with more spatial detail and larger spatial coverage would assist this effort.

The second of these challenges can be at least partially addressed through vigilant control of IPS, and through strong networks for communication among managers about IPS and new problems as they arise. Many of these networks exist in some form and are growing. Examples in the United States include Cooperative Weed Management Areas and Exotic Pest Plant Councils. Ideally, managers would address the challenge of increasing invasiveness of some species by controlling all unwanted nonnatives, but this approach is not practical in most locations.

Managers will certainly welcome decreases in competitiveness of IPS where they occur, and these cases will provide restoration opportunities for managers. The associated challenge will be learning how best to restore these areas, given the new climatic regime. Can the resident native species fill the gap created by the declining IPS, or should other, more climatically suitable species be used? The topic of "managed relocation," or the practice of relocating desirable species to areas that are newly climatically suitable, has generated lively debate (e.g., Hoegh-Guldberg et al., 2008; Ricciardi and Simberloff, 2009; Richardson et al., 2009). (The practice is also called "assisted colonization" and "assisted migration.") Ethical and legal concerns surrounding this practice will add to the challenge faced by managers.

Decreasing competitiveness of resident communities will provide a range of challenges for managers well beyond the scope of this book. However, to the extent this occurs it will presumably increase invasibility of the plant communities. Most of the debate on assisted migration centers around protecting the species that are most vulnerable to climate change, but it is also worth thinking about the degree to which such activities could or should be used to minimize susceptibility of landscapes to future conversion by IPS.

As we have demonstrated in this chapter, many biological interactions will affect the colonization, spread, and management of IPS as a function of global climate change. In the next chapter, we will examine a slightly different aspect of the harm that weeds do by drawing attention to the links between climate change, weeds, and public health.

References

Bais HP, Vepachedu R, Gilroy S, Callaway RM, Vivanco JM. 2003. Allelopathy and exotic plant invasion: from molecules and genes to species interactions. *Science* **301**:1377–1380.

Barney JN, DiTomaso JM. 2008. Nonnative species and bioenergy: are we cultivating the next invader? *Bioscience* **58**:64–70.

Beaumont LJ, Gallagher RV, Thuiller W, Downey PO, Leishman MR, Hughes L. 2009. Different climatic envelopes among invasive populations may lead to underestimations of current and future biological invasions. *Divers. Distrib.* **15**:409–420.

Beerling DJ, Huntley B, Bailey JP. 1995. Climate and the distribution of *Fallopia japonica*: use of an introduced species to test the predictive capacity of response surfaces. *J. Veg. Sci.* **6**:269–282.

Blair AC, Weston LA, Nissen SJ, Brunk GR, Hufbauer RA. 2009. The importance of analytical techniques in allelopathy studies with the reported allelochemical catechin as an example. *Biol. Inv.* **11**:325–332.

Blossey B, Notzold R. 1995. Evolution of increased competitive ability in invasive nonindigenous plants—a hypothesis. *J. Ecol.* **83**:887–889.

Blumenthal D., Chimner RA, Welker JM, Morgan JA. 2008. Increased snow facilitates plant invasion in mixedgrass prairie. *New Phytol.* **179**:440–448.

Blumenthal D, Mitchell CE, Pysek P, Jarosik V. 2009. Synergy between pathogen release and resource availability in plant invasion. *Proc. Natl. Acad. Sci. U. S. A.* **106**:7899–7904.

Bradley BA. 2009. Regional analysis of the impacts of climate change on cheatgrass invasion shows potential risk and opportunity. *Global Change Biol.* **15**:196–208.

Bradley BA, Mustard JF. 2005. Identifying land cover variability distinct from land cover change: cheatgrass in the Great Basin. *Remote Sens. Environ.* **94**:204–213.

Bradley BA, Oppenheimer M, Wilcove DS. 2009. Climate change and plant invasions: restoration opportunities ahead? *Global Change Biol.* **15**:1511–1521.

Brooks ML. 2003. Effects of increased soil nitrogen on the dominance of alien annual plants in the Mojave Desert. *J. Appl. Ecol.* **40**:344–353.

Brooks ML, D'Antonio CM, Richardson DM, Grace JB, Keeley JE, DiTomaso JM, Hobbs RJ, Pellant M, Pyke D. 2004. Effects of invasive alien plants on fire regimes. *Biosciences* **54**:677–688.

Byers JE. 2002. Impact of non-indigenous species on natives enhanced by anthropogenic alteration of selection regimes. *Oikos* **97**:449–458.

Callaway RM, Aschehoug ET. 2000. Invasive plants versus their new and old neighbors: a mechanism for exotic invasion. *Science* **290**:521–523.

Callaway RM, Ridenour WM. 2004. Novel weapons: invasive success and the evolution of increased competitive ability. *Front. Ecol. Environ.* **2**:436–443.

Chown SL, Slabber S, McGeoch MA, Janion C, Leinaas HP. 2007. Phenotypic plasticity mediates climate change responses among invasive and indigenous arthropods. *Proc. R. Soc. B-Biol. Sci.* **274**:2531–2537.

Colautti RI, MacIsaac HJ. 2004. A neutral terminology to define "invasive" species. *Divers. Distrib.* **10**:135–141.

Cotrufo MF, Ineson P, Scott A. 1998. Elevated CO_2 reduces the nitrogen concentration of plant tissues. *Global Change Biol.* **4**:43–54.

D'Antonio CM, Dudley T, Mack M. 2000. Disturbance and biological invasions: direct effects and feedbacks. In: Walker LR, (ed.) *Ecosystems of Disturbed Ground*. Elsevier, New York, pp. 429–468.

D'Antonio CM, Vitousek PM. 1992. Biological invasions by exotic grasses, the grass/fire cycle, and global change. *Annu. Rev. Ecol. Syst.* **23**:63–87.

Daehler CC. 2003. Performance comparisons of co-occurring native and alien invasive plants: Implications for conservation and restoration. *Ann. Rev. Ecol. Evol. Syst.* **34**:183–211.

Davis MA, Grime JP, Thompson K. 2000. Fluctuating resources in plant communities: a general theory of invasibility. *J. Ecol.* **88**:528–534.

Duke SO, Blair AC, Dayan FE, Johnson RD, Meepagala KM, Cook D, Bajsa J. 2009. Is (−)-catechin a novel weapon of spotted knapweed (*Centaurea stoebe*)? *J. Chem. Ecol.* **35**:141–153.

Dukes JS. 2011. Climate change. In: Simberloff D and Rejmánek M (eds.), *Encyclopedia of Biological Invasions*. University of California Press, Berkeley, CA, pp. 113–117.

Dukes JS, Mooney HA. 1999. Does global change increase the success of biological invaders? *TREE* **14**:135–139.

Dukes JS, Mooney HA. 2004. Disruption of ecosystem processes in western North America by invasive species. *Rev. Chil. Hist. Nat.* **77**:411–437.

Duncan CA, Jachetta JJ, Brown ML, Carrithers VF, Clark JK, DiTomaso JM, Lym RG, McDaniel KC, Renz MJ, Rice PM. 2004. Assessing the economic, environmental, and societal losses from invasive plants on rangeland and wildlands. *Weed Tech.* **18**:1411–1416.

Dyer AR, Rice KJ. 1999. Effects of competition on resource availability and growth of a California bunchgrass. *Ecology* **80**:2697–2710.

Ehrenfeld JG. 2003. Effects of exotic plant invasions on soil nutrient cycling processes. *Ecosystems* **6**:503–523.

Elton CS. 1958. *The Ecology of Invasions by Animals and Plants*. Chapman and Hall, Ltd., London.

Engelkes T, Morrien E, Verhoeven KJF, Bezemer TM, Biere A, Harvey JA, McIntyre LM, Tamis WLM, Van Der Putten WH. 2008. Successful range-expanding plants experience less above-ground and below-ground enemy impact. *Nature* **456**:946–948.

Galloway JN, Townsend AR, Erisman JW, Bekunda M, Cai ZC, Freney JC, Martinelli LA, Seitzinger SP, Sutton MA. 2008. Transformation of the nitrogen cycle: recent trends, questions, and potential solutions. *Science* **320**:889–892.

Genton BJ, Kotanen PM, Cheptou PO, Adolphe C, Shykoff JA. 2005. Enemy release but no evolutionary loss of defence in a plant invasion: an inter-continental reciprocal transplant experiment. *Oecologica* **146**:404–414.

Goodwin BJ, McAllister AJ, Fahrig L. 1999. Predicting invasiveness of plant species based on biological information. *Conserv. Biol.* **13**:422–426.

Gruber N, Galloway JN. 2008. An Earth-system perspective of the global nitrogen cycle. *Nature* **451**:293–296.

Gurevitch J, Padilla DK. 2004. Are invasive species a major cause of extinctions? *TREE* **19**:470–474.

Hoegh-Guldberg O, Hughes L, McIntyre S, Lindenmayer DB, Parmesan C, Possingham HP, Thomas CD. 2008. Assisted colonization and rapid climate change. *Science* **321**:345–346.

Huenneke LF, Hamburg SP, Koide R, Mooney HA, Vitousek PM. 1990. Effects of soil resources on plant invasion and community structure in Californian [USA] serpentine grassland. *Ecology* **71**:478–491.

IPCC. 2007. *Climate Change 2007: The Physical Science Basis*. Cambridge University Press, Cambridge, UK.

Jarnevich CS, Stohlgren TJ. 2009. Near term climate projections for invasive species distributions. *Biol. Invasions* **11**:1373–1379.

Joshi J, Vrieling K. 2005. The enemy release and EICA hypothesis revisited: incorporating the fundamental difference between specialist and generalist herbivores. *Ecol. Lett.* **8**:704–714.

Karl TR, Melillo JM, Peterson TC (eds). 2009. *Global Climate Change Impacts in the United States*. Cambridge University Press, Cambridge.

Keane RM, Crawley MJ. 2002. Exotic plant invasions and the enemy release hypothesis. *TREE* **17**:164–170.

Knapp AK, Beier C, Briske DD, Classen AT, Luo Y, Reichstein M, Smith MD, Smith SD, Bell JE, Fay PA, Heisler JL, Leavitt SW, Sherry R, Smith B, Weng E. 2008. Consequences of more extreme precipitation regimes for terrestrial ecosystems. *Biosciences* **58**:811–821.

Knapp PA. 1996. Cheatgrass (*Bromus tectorum* L) dominance in the Great Basin Desert: history, persistence, and influences to human activities. *Global Environ. Change* **6**:37–52.

Kriticos DJ, Randall RP. 2001. A comparison of systems to analyse potential weed distributions. In: Groves RH, Panetta FD, Virture JG (eds) *Weed Risk Assessment*. CSIRO Publishing, Melbourne, Australia, pp. 61–79.

Kriticos DJ, Sutherst RW, Brown JR, Adkins SW, Maywald GF. 2003a. Climate change and biotic invasions: a case history of a tropical woody vine. *Biol. Invasions* **5**:145–165.

Kriticos DJ, Sutherst RW, Brown JR, Adkins SW, Maywald GF. 2003b. Climate change and the potential distribution of an invasive alien plant: *Acacia nilotica* ssp *indica* in Australia. *J. Appl. Ecol.* **40**:111–124.

Lawler JJ, Tear TH, Pyke C, Shaw MR, Gonzalez P, Kareiva P, Hansen L, Hannah L, Klausmeyer K, Aldous A, Bienz C, Pearsall S. 2010. Resource management in a changing and uncertain climate. *Front. Ecol. Environ.* **8**:35–43.

Lee H, Reusser DA, Olden JD, Smith SS, Graham J, Burkett V, Dukes JS, Piorkowski RJ, McPhedran J. 2008. Integrated monitoring and information systems for managing aquatic invasive species in a changing climate. *Conserv. Biol.* **22**:575–584.

Levine JM, Vilá M, D'Antonio CM, Dukes JS, Grigulis K, Lavorel S. 2003. Mechanisms underlying the impacts of exotic plant invasions. *Proc. R. Soc. London B* **270**:775–781.

Lonsdale WM, Lane AM. 1994. Tourist vehicles as vectors of weed seeds in Kakadu National Park, northern Australia. *Biol. Conserv.* **69**:277–283.

Malcolm JR, Markham A, Neilson RP, Garaci M. 2002. Estimated migration rates under scenarios of global climate change. *J. Biogeogr.* **29**:835–849.

McFadyen REC. 1998. Biological control of weeds. *Annu. Rev. Entomol.* **43**:369–393.

Memmott J, Craze PG, Waser NM, Price MV. 2007. Global warming and the disruption of plant–pollinator interactions. *Ecol. Lett.* **10**:710–717.

Milchunas DG, Lauenroth WK. 1995. Inertia in plant community structure: state changes after cessation of nutrient-enrichment stress. *Ecol. Appl.* **5**:452–458.

Parker IM, Simberloff D, Lonsdale WM, Goodell K, Wonham M, Kareiva PM, Williamson MH, Von Holle B, Moyle PB, Byers JE, Goldwasser L. 1999. Impact: toward a framework for understanding ecological effects of invaders. *Biol. Inv.* **1**:3–19.

Patterson DT, Flint EP. 1990. Implications of increasing carbon dioxide and climate change for plant communities and competition in natural and managed ecosystems. In: Kimball BA, Rosenburg NJ, Allen LH Jr. (eds) *Impact of Carbon Dioxide, Trace Gases and Climate Change on Global Agriculture*. American Society of Agronomy, Madison, WI.

Pimentel D, Zuniga R, Morrison D. 2005. Update on the environmental and economic costs associated with alien-invasive species in the United States. *Ecol. Econ.* **52**:273–288.

Poorter H, Navas M-L. 2003. Plant growth and competition at elevated CO_2: on winners, losers and functional groups. *New Phytol.* **157**:175–198.

Poorter H, Roumet C, Campbell BD. 1996. Interspecific variation in the growth response of plants to elevated CO_2: a search for functional types. In: Körner C, Bazzaz FA (eds) *Carbon Dioxide, Populations, and Communities*. Academic Press, San Diego, CA, pp. 375–412.

Prati D, Bossdorf O. 2004. Allelopathic inhibition of germination by *Alliaria petiolata* (Brassicaceae). *Am. J. Bot.* **91**:285–288.

Prentiss AN. 1889. On root propagation of Canada thistle. *Cornell Agr. Expt. Sta. Bul.* **15**:190–192.

Pyšek P, Richardson DM. 2007. Traits associated with invasiveness in alien plants: where do we stand? In: Nentwig W (ed.) *Biological Invasions*. Springer-Verlag, Berlin, pp. 97–125.

Qian H, Ricklefs RE. 2006. The role of exotic species in homogenizing the North American flora. *Ecol. Lett.* **9**:1293–1298.

Raghu S, Anderson RC, Daehler CC, Davis AS, Wiedenmann RN, Simberloff D, Mack RN. 2006. Adding biofuels to the invasive species fire? *Science* **313**:1742–1742.

Rejmánek M. 1995. What makes a species invasive? In: Pysek P, Prach K, Rejmánek M, Wade M (eds) *Plant Invasions: General Aspects and Special Problems*. SPB Academic Publishing, Amsterdam, pp. 3–13.

Rejmánek M, Richardson DM. 1996. What attributes make some plant species more invasive? *Ecology* **77**:1655–1661.

Ricciardi A, Simberloff D. 2009. Assisted colonization is not a viable conservation strategy. *TREE* **24**:248–253.

Rice PM. 2005. Downy brome, *Bromus tectorum* L. In: Duncan CA and Clark JK (eds.), *Invasive Plants of Range and Wildlands and Their Environmental, Economic, and Societal Impacts*. Lawrence, KS: Weed Science Society of America, pp. 147–170.

Richardson DM, Allsopp N, D'Antonio CM, Milton SJ, Rejmanek M. 2000a. Plant invasions—the role of mutualisms. *Biol. Rev.* **75**:65–93.

Richardson DM, Bond WJ, Dean RJ, Higgins SI, Midgley GF, Milton SJ, Powrie LW, Rutherford MC, Samways MJ, Schulze RE. 2000b. Invasive alien species and global change: a South African perspective. In: Mooney HA, Hobbs RJ (eds) *Invasive Species in a Changing World.* Island Press, Washington, pp. 303–350.

Richardson DM, Hellmann JJ, McLachlan JS, Sax DF, Schwartz MW, Gonzalez P, Brennan EJ, Camacho A, Root TL, Sala OE, Schneider SH, Ashe DM, Clark JR, Early R, Etterson JR, Fielder ED, Gill JL, Minteer BA, Polasky S, Safford HD, Thompson AR, Vellend M. 2009. Multidimensional evaluation of managed relocation. *Proc. Natl. Acad. Sci. U. S. A.* **106**:9721–9724.

Scherer-Lorenzen M, Olde Venterink H, Buschmann H. 2007. Nitrogen enrichment and plant invasions: the importance of nitrogen-fixing plants and anthropogenic eutrophication. In: Nentwig W (ed.) *Biological Invasions*. Springer-Verlag, Berlin, pp. 163–180.

Serbesoff-King, K. 2003. Melaleuca in Florida: a literature review on the taxonomy, distribution, biology, ecology, economic importance and control measures. *J. Aquat. Plant Manage.* **41**:98–112.

Shafroth PB, Cleverly JR, Dudley TL, Taylor JP, Van Riper C, Weeks EP, Stuart JN. 2005. Control of Tamarix in the Western United States: implications for water salvage, wildlife use, and riparian restoration. *Environ. Manage.* **35**:231–246.

Shea K, Chesson P. 2002. Community ecology theory as a framework for biological invasions. *TREE* **17**:170–176.

Skinner K, Smith L, Rice P. 2000. Using noxious weed lists to prioritize targets for developing weed management strategies. *Weed Sci.* **48**:640–644.

Smith SD, Huxman TE, Zitzer SF, Charlet TN, Housman DC, Coleman JS, Fenstermaker LK, Seemann JR, Nowak RS. 2000. Elevated CO_2 increases productivity and invasive species success in an arid ecosystem. *Nature* **408**:79–82.

Smith SD, Strain BR, Sharkey TD. 1987. Effects of carbon dioxide enrichment on four Great Basin [USA] grasses. *Funct. Ecol.* **1**:139–144.

Song LY, Wu JR, Li CH, Li FR, Peng SL, Chen BM. 2009. Different responses of invasive and native species to elevated CO_2 concentration. *Acta Oecologica* **35**:128–135.

Stewart G, Hull AC. 1949. Cheatgrass (*Bromus tectorum* L): an ecologic intruder in southern Idaho. *Ecology* **30**:58–74.

Stinson KA, Campbell SA, Powell JR, Wolfe BE, Callaway RM, Thelen GC, Hallett SG, Prati D, Klironomos JN. 2006. Invasive plant suppresses the growth of native tree seedlings by disrupting belowground mutualisms. *PLoS Biol.* **4**:727–731.

Sutherst RW, Floyd RB, Maywald GF. 1995. The potential geographical distribution of the cane toad, *Bufo marinus* L. in Australia. *Conserv. Biol.* **9**:294–299.

Theoharides KA, Dukes JS. 2007. Plant invasion across space and time: factors affecting nonindigenous species success during four stages of invasion. *New Phytol.* **176**:256–273.

U.S. Congress, O.T.A. 1993. *Harmful Non-Indigenous Species in the United States.* U.S. Government Printing Office, Washington, DC.

Vermeij GJ. 1996. An agenda for invasion biology. *Biol. Conserv.* **78**:3–9.

Vivanco JM, Bais HP, Stermitz FR, Thelen GC, Callaway RM. 2004. Biogeographical variation in community response to root allelochemistry: novel weapons and exotic invasion. *Ecol. Lett.* **7**:285–292.

Wedin DA, Tilman D. 1996. Influence of nitrogen loading and species composition on the carbon balance of grasslands. *Science* **274**:1720–1723.

Weeks EP, Weaver HL, Campbell GS, Tanner BD. 1987. *Water Use by Saltcedar and by Replacement Vegetation in the Pecos River Floodplain between Acme and Artesia.* U.S. Geological Survey Professional Paper 491-G, New Mexico, 33 pp.

Weiss SB. 2000. Cars, cows, and checkerspot butterflies: nitrogen deposition and management of nutrient poor grasslands for a threatened species. *Conserv. Biol.* **13**:1476–1486.

Whisenant SG. 1990. Changing fire frequencies on Idaho's Snake River plains: ecological and management implications. In: McArthur ED, Romney EM, Smith SD, Tueller PT (eds) *Proceedings—Symposium on Cheatgrass Invasion, Shrub Die-off, and Other Aspects of Shrub Biology and Management*. Intermountain Research Station, U.S. Forest Service, Ogden, Utah, pp. 4–10.

Wiedemann AM, Pickart A. 1996. The *Ammophila* problem on the northwest coast of North America. *Landscape Urban Plann.* **34**:287–299.

Zavaleta E. 2000. Valuing ecosystem services lost to *Tamarix* invasion in the United States. In: Mooney HA, Hobbs RJ (eds) *Invasive Species in a Changing World.* Island Press, Washington, pp. 261–300.

Ziska LH. 2003a. Evaluation of the growth response of six invasive species to past, present and future atmospheric carbon dioxide. *J. Exp. Bot.* **54**:395–404.

Ziska LH. 2003b. The impact of nitrogen supply on the potential response of a noxious, invasive weed, Canada thistle (*Cirsium arvense*) to recent increases in atmospheric carbon dioxide. *Physiol. Plant.* **119**:105–112.

Ziska LH, Reeves JB, Blank R. 2005. The impact of recent increases in atmospheric CO_2 on biomass production and vegetative retention of Cheatgrass (*Bromus tectorum*): implications for fire disturbance. *Global Change Biol.* **11**:1325–1332.

7 Weeds, CO_2, Climate, and Health

"Within the infant rind of this small flower Poison lives, and medicine power."
William Shakespeare

Introduction

We have thus far, tried to provide a sense of the magnitude and extent of the damage inflicted by weed species, not only for agricultural systems, but also as we emphasized in the last chapter, IPS in unmanaged, natural plant communities. We have seen numerous examples of how increasing CO_2 and climate change are likely to alter the impact of both agricultural and invasive weed species (IPS), and some of the potential consequences.

In this chapter, we will attempt to expand our appreciation of the harm that weeds do by focusing on an underappreciated but very relevant aspect of weed biology: the effects on human health. Although many weed scientists tend to focus on the agronomic and environmental threats posed by weeds, the general public is very cognizant of how weeds can directly affect their health: through mechanical injury, allergenic reactions, contact dermatitis, or toxicology/pharmacology. There are also indirect means by which weeds can alter health, including nutrition and potential expansion of animal disease reservoirs. Our objective in the current chapter is to review these links between weed biology and health and to assess how CO_2 and climate are likely to alter the biological associations and determine, if possible, the likely consequences for human or animal physiology.

Direct Effects

Mechanical Injury

Many weeds have spines, burs, needles, thorns, or other appendages that can puncture the skin of the unwary. Weed-pullers whose bare skin bumps into Canada thistle or stinging nettle will be particularly unhappy. It is not uncommon for barefooted children to step on sandbur or puncturevine fruits. Cuts and puncture wounds hurt, and can get infected without proper treatment.

Some weeds also wound domestic livestock because of thorns, barbs, stiff hairs, or sharply pointed seeds. Wounds to eyes, mouth parts, skin, or the digestive system can lower the commercial value of meat, wool, or hides. In some cases, open wounds may invite attack from parasites such as screw worms (the larva of certain flies of the genus *Callitroga*) which can kill the animal.

One of the few studies to quantify responses of plant mechanical protections to elevated CO_2 or climate showed an increase in average spine length for immature leaves of Canada thistle. Plants grown above a 285-ppm baseline also had more spines per mature leaf (Ziska, 2002; Figure 7.1).

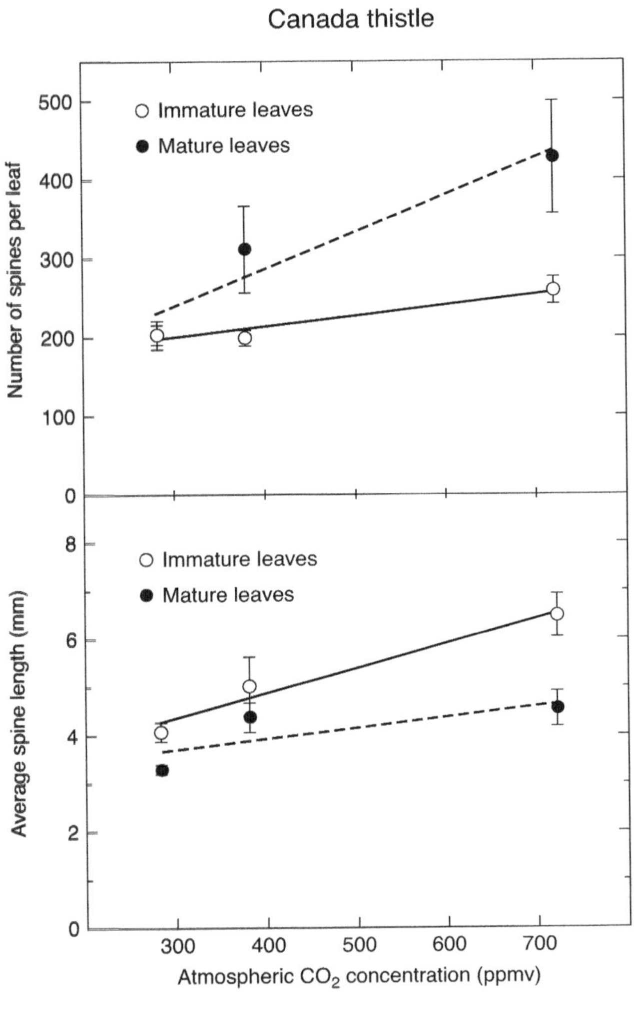

Figure 7.1 Changes in spine length and number for mature (fully expanded) and immature leaves of Canada thistle as a function of recent (285–380 ppm) and projected (720 ppm) increases in atmospheric carbon dioxide (see Ziska, 2002 for details).

Aerobiology

Allergic diseases are ranked as the sixth leading cause of chronic illness in the United States, affecting approximately 17% of the population (AAAAI, 2000). The Centers for Disease Control (CDC) estimate asthma prevalence at about 25 million individuals (9 million children, 16 million adults; CDC, 2002). Self-reporting of asthma appears to be on the rise, with the largest increase occurring in preschool-aged children (Mannino et al., 1998; Figure 7.2).

Although the epidemiology of allergic diseases like asthma is complex, (e.g., genetic predispositions and access to health care), exposure to allergens from the natural environment, such as pollen grains or fungal spores is unequivocally associated with the symptomology and acceleration of human respiratory disease (Frenz, 2001). Recently, the *Journal of Allergy and Clinical Immunology* (Shea et al., 2008) stated that "there is now a wealth of evidence that climate change has had, and will have, further impact on a variety of allergenic plants."

Plant induced increases in allergic rhinitis (i.e., hay fever) and asthma are associated with three distinct seasonal sources of plant-based allergens: trees, grasses, and ragweed for the spring, summer, and fall seasons, respectively. Although projected changes in trees and grasses with CO_2 and climate are interesting in their own right (see Ziska et al., 2008a), the most information is currently available regarding ragweeds (plants in the *Ambrosia* genus).

Figure 7.2 Sampling pollen from common ragweed grown under different CO_2 environments.

Initial experimental chamber or greenhouse studies indicated that common ragweed could respond to projected CO_2 end of century concentrations (Wayne et al., 2002; Ziska and Caulfield, 2000). In addition, similar studies have shown that ragweed growth and pollen production is likely to have responded both quantitatively and qualitatively to the increase in CO_2 concentration that has occurred during the twentieth century (Singer et al., 2005). For western ragweed, simulated increases in temperature (+4°C) during the summer resulted in increased growth as well as regrowth following cutting with a subsequent increase of ~85% in overall pollen production (Wan et al., 2002). Additional experimentation demonstrated that earlier seed germination in common ragweed as a result of simultaneous increases in CO_2 and warming also stimulated growth and increased catkin production in climate-controlled glasshouses (Rogers et al., 2006).

In some ways, the environment created by cities may provide an analog for studying ecological responses to global change (Carreiro and Tripler, 2005). Urban levels of carbon dioxide and air temperature as well as season length (number of frost-free days) can be significantly greater. Ziska and colleagues utilized an existing macro-climate gradient between Baltimore and the surrounding Maryland counties to examine the quantitative aspects of ragweed growth and pollen production in 2000 and 2001. Overall, ragweed from the same seed source, and from the same soil but grown along this transect showed urban-induced earlier flowering and greater pollen production when compared to suburban or rural locations (Ziska et al., 2003). While it remains unclear whether urbanization would consistently result in similar responses either spatially (i.e., other cities), or temporally (sustainability of increased pollen production) in regard to ragweed biology, use of an urban–rural macro-climate and the associated higher CO_2 concentrations and air temperatures may be a harbinger of what could be expected globally with pollen production and allergic rhinitis/asthma and global climate change (Figure 7.3).

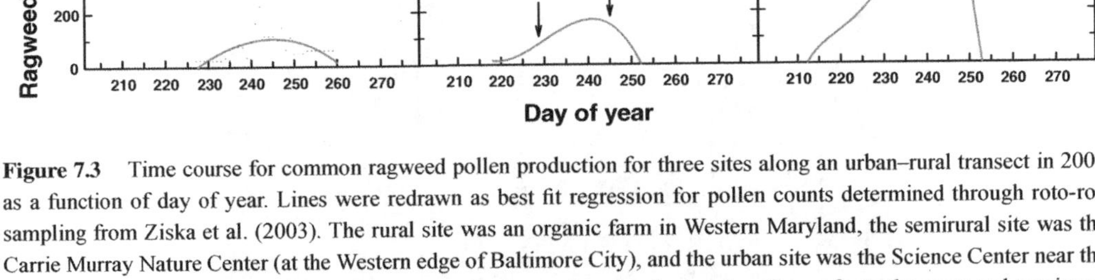

Figure 7.3 Time course for common ragweed pollen production for three sites along an urban–rural transect in 2001 as a function of day of year. Lines were redrawn as best fit regression for pollen counts determined through roto-rod sampling from Ziska et al. (2003). The rural site was an organic farm in Western Maryland, the semirural site was the Carrie Murray Nature Center (at the Western edge of Baltimore City), and the urban site was the Science Center near the Inner Harbor in downtown Baltimore. The arrows for the semirural and urban locations refer to the onset and maximum pollen dates for the rural site.

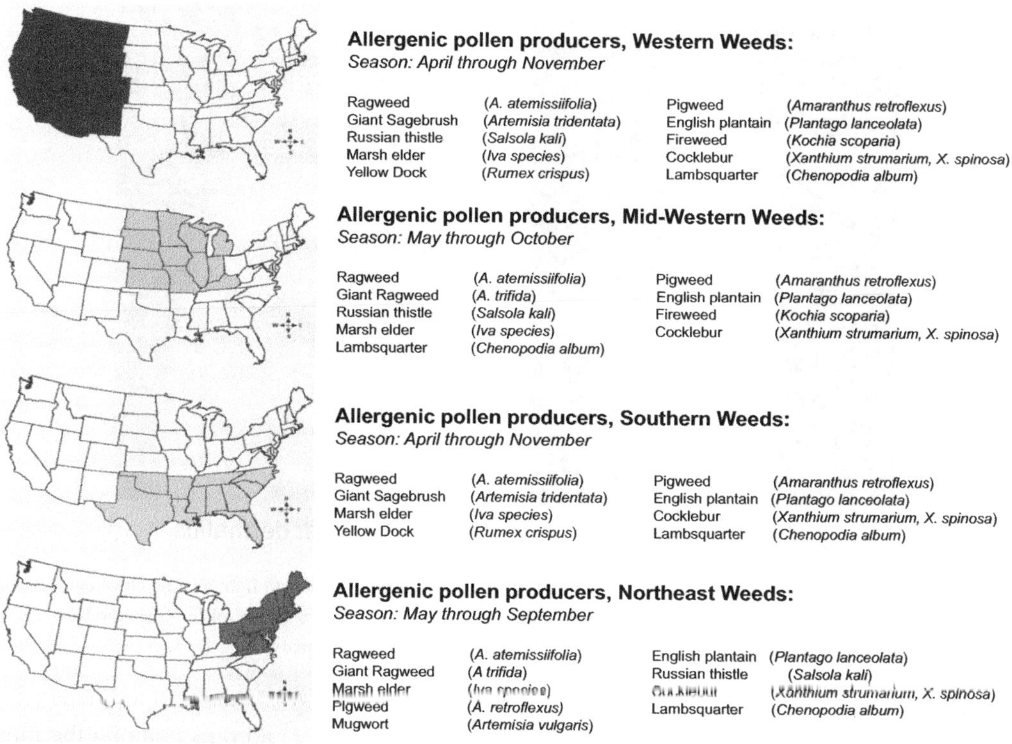

Figure 7.4 List of allergenic weeds by region within the United States. Rising carbon dioxide and climate change may alter the range and pollen production of these weeds.

There are several other known allergenic weeds, including lambsquarters, cocklebur, marsh elder, mugwort, plantain, etc. For at least some of these (e.g., lambsquarters), the response to CO_2 has been documented (Patterson, 1995); however, few studies have quantified pollen production or allergenicity for these weeds in response to either rising CO_2 or climate-induced changes in seasonality. While additional data are needed, the data to date suggest climate forcing and rising CO_2 (both at the local urban level, and that projected globally) have the potential to affect allergen production of weed species (Figure 7.4).

Dermatitis

Dermatitis (an autoimmune reaction from dermal contact) can occur as photo dermatitis (increased sensitivity to light), for example, giant hogweed, but perhaps the most widely recognized are those that result in contact dermatitis, for example, the poison ivy group (*Toxicodendron* spp.). As with weeds and aerobiology, only a handful of studies have examined the impact of rising CO_2 and/or climate on changes in weed-induced dermatitis, with most of the focus on poison ivy (Mohan et al., 2006; Ziska et al., 2007; Figure 7.5).

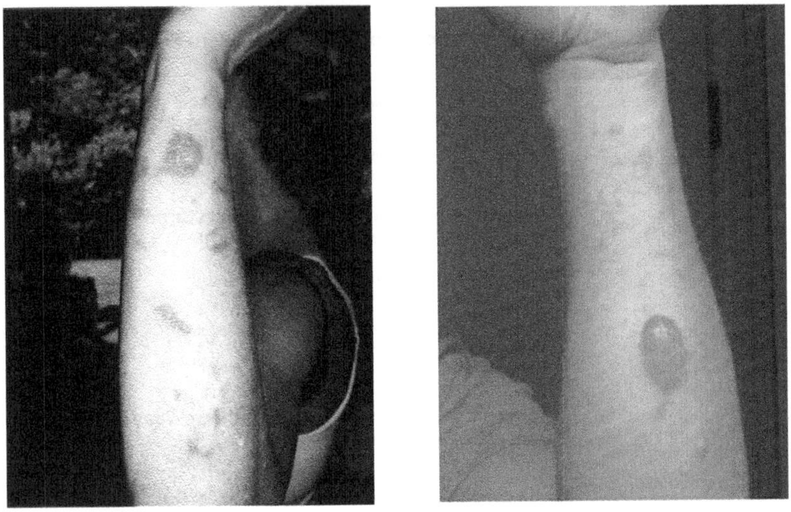

Photo dermatitis Contact dermatitis

Figure 7.5 Examples of dermatitis: on the left, increased photo-sensitivity of skin to light as a result of contact with giant hogweed; on the right, contact dermatitis from poison ivy. (Photo on left is courtesy of Dr. Suzanne Rowe, King County Noxious Weed Program, Seattle, Washington; Photo on right is from Wikipedia.)

Poison ivy, while native to North America, is found globally and ranks among the most medically problematic weeds. It is estimated that 25–40 million people in the United States require medical treatment annually after exposure (Baer, 1990). In a 6-year study conducted at the Duke University Free-Air CO_2 Enrichment (FACE) facility, Mohan et al. (2006) reported that elevated levels of CO_2 (+200 ppm above ambient) increased photosynthesis, water use efficiency, growth, and population biomass of poison ivy in an intact forest ecosystem. In addition, high-CO_2-grown plants produced a more allergenic form of urushiol. Additional studies with poison ivy (Ziska et al., 2007) grown at 100 ppm CO_2 increments (300, 400, 500, and 600 ppm), for a 250-day period found CO_2-induced changes in leaf area, leaf and stem weight, and rhizome length. In contrast to the Mohan et al. study, there was a nonsignificant effect ($P = 0.18$) of rising CO_2 level on urushiol concentration; however, because of the stimulatory effect of CO_2 on biomass, the amount of urushiol per plant increased significantly for all CO_2 values above the 300 ppm baseline. In addition, significant increases in the rate of leaf development for poison ivy following leaf removal were also observed with rising CO_2, suggesting greater recovery from herbivory.

It is worthwhile to note that the CO_2 sensitivity of poison ivy growth is consistent with in situ observations for other vine species (e.g., Belote et al., 2004; Phillips et al., 2002). It has been suggested that one possible reason for the greater response among vine species is related to their ability to allocate more resources to additional leaf manufacture because of low allocation to woody (nonphotosynthesizing) support structures (Sasek and Strain, 1990; Tsugawa et al., 1980); greater light interception in turn, could, over time, confer greater competitive advantage. Overall, the current data for poison ivy do suggest a plausible link between rising carbon dioxide, weed biology, and increased dermatitis. At present, potential climatic interactions such as warmer temperatures and longer growing season in relation to poison ivy development and urushiol content are unknown.

Toxicology

Similar to dermatitis, the presence of toxic substances is related to specific plant organs (fruit, leaf, and stem), and edible and poisonous parts can exist on the same plant (e.g., rhubarb). There are over 700 plant species, including a number of weed species, that are poisonous to humans and livestock. Chemicals form a principal defense against herbivory in a number of agronomic and generalist weeds including castor bean, jimson weed, rosary pea, water hemlock, and common nightshade, inter alia. The chemicals in these weeds range from ricin and abrin, phytotoxins found in castor bean and rosary pea; atropine and scopolamine, strong anticholinergics found in jimson weed and nightshade, to coniine, a potent neurotoxin found in water hemlock. Many of these compounds are sufficiently lethal that death can occur in small doses (e.g., ricin is sufficiently lethal as to be considered a weapon of mass destruction by the U.S. military).

Ingestion of plant material continues to be a very common exposure for humans (particularly children), and can account for nearly 100,000 calls to national poison centers annually (Watson et al., 2004). Pediatric patients constitute more than 80% of plant-related exposures; however, less than 20% of plant exposures require medical treatment (Watson et al., 2004). For livestock, a partial list of weeds and their poisons would include buttercups (protoanemonin), horsetail (thiaminase), pigweeds (oxalate and nitrate), pokeweed (oxalic acid), white snakeroot (glycosides), and yellow starthistle (various neurotoxins). The consequences of animal poisoning depend on the quantity of the toxin consumed, size and species of the animal, general health of the animal, and concentration of the toxin in the weed; the latter varying with stage of growth, soil conditions (especially the nutrient and moisture level), and biotype.

Only a handful of studies have determined the likely effect of rising CO_2 and/or climate change on poisonous weeds. Recent work by Vanaja et al. (2008) demonstrated a strong growth and yield response of castor bean to elevated CO_2 concentrations of 550 and 700 ppm. However, rising CO_2 did not affect either the content of quality of the castor oil. The impact of rising CO_2 concentration (300, 400, and 700 ppm) at two different temperatures (ambient and ambient $+5°C$) was also determined for two alkaloids in jimsonweed—atropine and scopolamine (Ziska et al., 2005). Elevated CO_2 had no effect on atropine concentration, but significantly increased the concentration of scopolamine. Conversely, temperature per se had no effect on scopolamine, but increased the concentration of atropine per plant. Other poisonous weeds have been examined in response to rising CO_2 (e.g., pokeweed; He et al., 2005), but specific toxicological evaluations have not been conducted. However, to date, these initial results suggest that CO_2 and/or climate could have a significant effect both on the growth of poisonous weeds, and their toxicology.

Indirect Effects

Air Quality

It has been recognized for a number of years that isoprene emissions from vascular plants play an important role in atmospheric chemistry. On warm, sunny days, isoprene can mix with nitrogen oxide to form ozone, a primary (and hazardous) constituent of smog. Kudzu, the invasive vine occupying about 8 million acres in the southeastern United States, produces these gases prolifically. Fast-growing kudzu leaves can emit isoprene into the air, and kudzu roots can convert atmospheric nitrogen into ammonium, some of which gets converted by soil bacteria into nitric oxide (Lerdau et al., 1997; Sharkey et al., 2005). Kudzu responds strongly to carbon dioxide in greenhouse conditions, and northern migration in response to global warming may also be occurring (Sasek and Strain, 1990).

Sometimes the links between plant biology and human illness are less than straightforward. Consider the case of Nancy Hanks, Abraham Lincoln's mother, who died in 1818 when "milk sickness" broke out in the Little Pigeon Creek Settlement in Indiana.

What exactly is "milk sickness"? White snakeroot is a shade-loving weed found throughout the Midwest. White snakeroot also contains the toxin tremetol, which is fat soluble and excreted in the milk of lactating animals. Milk sickness is what occurs when a person drinks the milk from cows that have eaten white snakeroot and pass on the toxin. It is also referred to as puking fever, sick stomach, the slows, and the trembles.

What does this have to do with climate? The illness was most common in dry years when cows wandered from poor pastures to the woods in search of food. Pasteurization and the mixing of milk from many different cows reduces the chance of contracting milk sickness today, but in the mid-nineteenth century the sickness was well known in the Midwest. Abraham Lincoln was only 9 when his mother passed away.

Narcotics

Among the most pernicious of weeds are those plants whose secondary products become illicit or addictive within a society. The abuse of such drugs (whether legal or illegal) carries a high economic and personal cost within a society. For the United States, the societal costs are reflected in health care expenses, work loss, property damage, prevention, treatment, incarceration, etc. All told, these costs can exceed $200 billion in a given year (ONDCP, 2001).

If there is then, a "war on drugs," it is of obvious importance to determine whether climate change and/or CO_2 can alter the growth and production of those weed species associated with this "war." Among the Papaveraceae family, the genus *Papaver* comprises about 110 species of both annual and perennial herbs; however, only *Papaver somniferum* and *Papaver setigerum* are known to produce opiates, with the concentration of these opiates varying considerably between species (e.g., Garnock-Jones and Scholes, 1990; La Valva et al., 1985). For *P. somniferum*, optimal development rates did not vary for day/night temperatures of 20/15, 24/19, and 28/23°C (Acock et al., 1997). For wild poppy (*P. setigerum*), growth over a range of carbon dioxide concentrations (300, 400, 500, and 600 ppm) had a significant effect on both leaf area and vegetative biomass (Ziska et al., 2008b). Carbon dioxide increases had no stimulatory effect on alkaloid concentration for codeine, papaverine, and noscapine; although significant increases in morphine concentration were observed. However, whole plant production of all opiates increased significantly in response to rising carbon dioxide because the stimulatory effect of CO_2 on capsule size and number compensated for any CO_2-induced change in concentration. In addition, the relative proportion of morphine (to the other alkaloids) increased significantly up to a CO_2 of 500 ppm (10.4, 11.7, 12.9, and 12.4%, respectively for the 300, 400, 500, and 600 ppm CO_2 treatments; Figure 7.6).

Although these data provide a basis for anticipating other CO_2 or climate-influenced changes in narcotic plants, specific data on species such as *Erythoxylem coca* (cocaine), *Salvia divinoum* (Salvinorin A), *Artemiesia absinthe* (absinthe), *Lophophora williamsii* (mescaline), etc., are entirely unknown. Ironically, there are a number of apocryphal data sites on the web for marijuana that purport to demonstrate the impact of CO_2, temperature, etc., on its physiology. It is unfortunate that no similar effort has been made on the part of scientists or health-care providers to understand the impact of CO_2 and climate change on the spread or potential efficacy of this or other narcotic plant species.

Nutrition

Some of the earliest studies that examined the impact of weeds on crop production documented the weed-induced changes in soil and crop nutrient status (e.g., Ince, 1915). In the 1950s, Vengris and colleagues observed how the presence or absence of weeds altered the nutritional concentration of corn when grown under optimal fertilizer (Vengris et al., 1955). They compared corn with no weeds with corn grown with redroot pigweed, common lambsquarters, crabgrass, and barnyardgrass, individually (all other weeds being removed). Weeds decreased nitrogen and potassium concentrations in corn; in contrast, the overall nutrient status of the individual weeds was unchanged by competition (Table 7.1).

Since these initial trials, numerous data have emerged that confirm the negative effect of weeds on nutritional resources, with weed species depleting nutrients as rapidly (or more rapidly) than many crop species. This has obvious implications for weed management (i.e., you cannot reduce the impact of weeds by adding more fertilizer), but it also demonstrates the negative effect weeds have on the nutritional quality of food. This effect has been observed in potato (Saghir and Markoullis, 1974) and in wheat (from white sweet clover, or wild

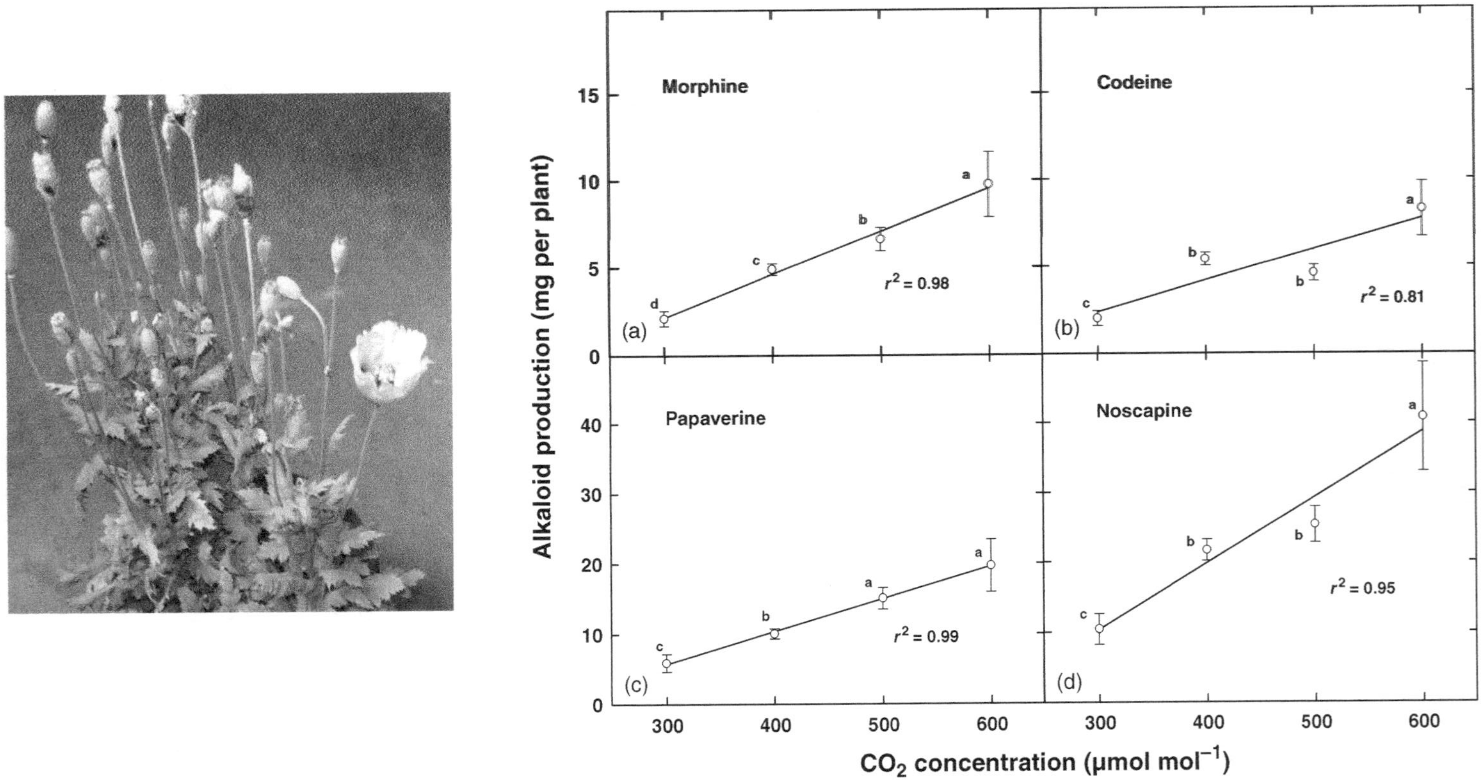

Figure 7.6 Change in the production of opiates from wild poppy as a function of rising CO_2 above a mid-twentieth century baseline (~300 ppm). (Photo of wild poppy courtesy of Ms. Sini Panicker of U.S. Drug Enforcement Agency.)

Table 7.1 Quantitative changes in percent of nitrogen, potassium, and calcium content in corn with and without weed competition as adapted from Vengris et al. (1955).

Plants present	Age	N%	K%	Ca%
Corn, alone	Young	3.78	4.02	0.40
Mature	1.44	0.90	0.17	
Corn and weeds	Young	3.05	3.08	0.42
Mature	1.38	0.70	0.20	

Samples were taken 1 month after emergence (young) and again at harvest (mature).

onion), where it becomes apparent through changes in the odor and baking qualities of flour. Similarly, weeds impair quality of animal produce, with weeds inducing undesirable flavors to milk and meat such as wild onion and wild garlic.

Clearly, because weeds affect nutritional quality of meat, diary, and crop products, they have an indirect effect on public health. Less clear however, is the role CO_2 and climate have in modifying such an effect. This is due in part to the direct effect of rising CO_2 on plant nutrient status and quality. For example, a study of spring wheat lines released during the twentieth century (when atmospheric carbon dioxide increased from ~300 to 380 ppm) found a decline in protein concentration and changes in flour quality (see Table 4 in Ziska et al., 2004). Several studies have found CO_2-induced declines in protein concentration for important cereal species (Ziska et al., 2004). In addition, some data suggest that rising CO_2 could decrease the micronutrient concentration of some crop species (Loladze, 2002). It has been recognized, of course, that year-to-year variation in climate can alter the nutritional content of food and fodder, but the direct effects of rising CO_2 and/or climate on nutritional quality have been less well characterized. This is unfortunate, since nutrition plays an important part in overall human health.

Weeds as a Facilitator of Disease Vectors

El Nino and the Hantavirus

In May of 1993, a young, healthy Navajo man was rushed to the emergency room of a hospital in New Mexico. Suffering from shortness of breath, the man died a short time after admission. Following his death, medical personnel found that the young man's fiancée had died just a few days previously, after showing similar symptoms.

The sudden deaths of two young healthy individuals led to large headlines, public interest, and a concerted effort on the part of health officials at the Indian Health Service (IHS), the New Mexico Office of Medical Investigations (OMI), and the Center for Disease Control (CDC) to find the cause of the disease. Soon, Dr. Bruce Tempest of IHS working with OMI had located an additional five healthy people who had all died after acute respiratory failure. After a number of false leads, tissue samples sent to CDC virologists indicated that the pulmonary symptoms were likely related to a virus, in particular a previously unknown strain of Hantavirus.

Researchers already knew that rodents are the principal carriers for Hantavirus, and a massive effort was launched to capture as many mice and rats in the area as possible. The results showed that approximately 30% of the deer mice were infected with the Hantavirus. By November, the specific Hantavirus associated with the outbreak had been isolated by the special pathogens branch at the CDC. The new virus was called the Sin Nombre virus, or SNV, and the subsequent disease, Hantavirus pulmonary syndrome, or HPS. By the start of 1994, over 55 HPS cases had been documented, 32 of them fatal.

But why the sudden outbreak of a previously obscure disease? The essential answer is that just prior to the epidemic there was a sudden, rapid surge in the deer mouse population. Why? The answer rests with the El Nino Southern Oscillation (ENSO) event that occurred in 1991–1992 and the subsequent warmer winter (1992–1993) and rainy spring (1993). These climatic conditions resulted in an explosive growth of vegetation, with a tenfold increase in the number of deer mice. With larger, unprecedented numbers of deer mice, contact with humans was exacerbated, and transmission of the virus occurred (CDC, 1996).

Weeds and Disease Vectors

How do weeds fit into an epidemiological model like that of Hantavirus transmission? While weeds do not transmit disease themselves, weeds can and do serve as food sources. Where and how much they grow therefore will influence where animal disease vectors will travel. For example, food sources for deer mice in New Mexico include a number of weedy species, such as spotted knapweed (e.g., Pearson and Ortega, 2001). Such an arrangement may be mutually beneficial as deer mice also distribute undigested spotted knapweed seeds.

Similarly, mosquitoes are well-known vectors for disease transmission. In particular, the malarial parasite (Plasmodium) carried by the Anopheles mosquito remains a leading cause of mortality, particularly in young children, with approximately 2 million deaths occurring annually according to the CDC. Other viral diseases carried by mosquitoes include yellow fever, dengue fever, Rift Valley fever, West Nile virus, eastern equine, and western equine encephalitis. Because of the high rate of mortality, the later two viruses are among the most serious mosquito-borne diseases in the United States. As with Hantavirus, plants may play a critical role in mosquito biology. For example, while female mosquitoes require blood for egg development, both male and female mosquitoes feed primarily on plant nectar. Pollen is also recognized as a food source for mosquito larvae (Ye-Ebiyo et al., 2000). As with deer mice and the Hantavirus, where and how much food plants provide will determine the spread and population density of a number of mosquito genera, with subsequent effects on human and animal disease.

As observed for the Hantavirus, the impact of ENSO-induced changes in temperature and precipitation are likely to have significant effects on the food supply of animal disease vectors. That food supply, in turn, is represented, at least in part, by a number of generalist weed species within a community. The extent to which climate and/or rising CO_2 will impact those species, both in regard to size, phenology, pollen production, etc., as well as their ecosystem characteristics, (e.g., range and distribution), will certainly have an effect on the number of disease vectors they can support. The role of weeds as facilitators of human disease does not, in general, appear to be well understood within the health care community. As such, weed biologists and plant scientists may play a unique part in elucidating such a role, particularly in the context of climatic change.

Summary

In the last few chapters, we have attempted to provide an overview of how weeds damage not only agricultural and natural systems, but also human health, and how climate and CO_2 are likely to modify the degree of damage. We recognize that the interactions between public health and weed biology in the current chapter may be somewhat outside the traditional viewpoint of many weed biologists. However, as we have hopefully shown, there are a number of overlapping research areas where unique and valuable contributions can be made by weed scientists. These areas range from understanding how climate and CO_2 can alter the direct effects of weeds on aerobiology and dermatitis, to indirect effects on nutrition and narcotics, as well as the potential role of weeds in facilitating the spread of animal vectors of disease. In the following chapters, we will turn our attention to ways in which the weed damage we have described can be modified or controlled, and how changing climate and rising carbon dioxide could modify the efficacy of different control strategies.

References

AAAAI. 2000. The Allergy Report. American Academy of Allergy Asthma and Immunology, Milwaukee, WI.

Acock MC, Pausch RC, Acock B. 1997. Growth and development of opium poppy (*Papaver somniferum* L.) as a function of temperature. *Biotronics* **26**:47–57.

Baer RL. 1990. Poison ivy dermatitis. *Cutis* **46**:34–36.

Belote RT, Weltzin JF, Norby RJ. 2004. Response of an understory plant community to elevated CO_2 depends on differential responses of dominant invasive species and is mediated by soil water availability. *New Phytol.* **161**:827–835.

Carreiro MM, Tripler CE. 2005. Forest remnants along urban rural gradients: examining their potential for global change research. *Ecosystems* **8**:568–582.

CDC (Centers for Disease Control). 1996. Hantavirus pulmonary syndrome—United States, 1995 and 1996. *MMWR Morb. Mortal. Wkly. Rep.* **45**:291–295.

CDC (Centers for Disease Control). 2002. Asthma prevalence and control characteristics by race/ethnicity—United States. *MMWR Morb. Mortal. Wkly. Rep.* **53**:145–148.

Frenz DA. 2001. Interpreting atmospheric pollen counts for use in clinical allergy: allergic symptomology. *Ann. Allergy. Asthma Immunol.* **86**:150–158.

Garnock-Jones PJ, Scholes P. 1990. Alkaloid content of *Papaver somniferum* subsp. *setigerum* from New Zealand. *N. Z. J. Bot.* **28**:367–369.

He J-S, Wolfe-Bellin KS, Bazzaz FA. 2005. Leaf level physiology, biomass, and reproduction of *Phytolacca americana* under conditions of elevated CO_2 and altered temperature regimes. *Int. J. Plant Sci.* **166**:615–622.

Ince JW. 1915. Fertility and weeds. *North Dakota Agric. Bull.* **112**:233–247.

La Valva V, Sabato S, Gigliano GS. 1985. Morphology and alkaloid chemistry of *Papaver setigerum* D.C. (Papaveraceae). *Taxon* **34**:191–196.

Lerdau M, Guenther A, Monson R. 1997. Plant production and emission of volatile organic compounds. *Biosciences* **47**:373–383.

Loladze I. 2002. Rising atmospheric CO_2 and human nutrition: toward globally imbalanced plant stoichiometry? *TREE* **17**:457–461.

Mannino DM, Homa DM, Pertowski CA. 1998. Surveillance for asthma prevalence—United States, 1960–1995. *MMWR Morb. Mortal. Wkly. Rep.* **47**(SS-1):1–28.

Mohan JE, Ziska LH, Schlesinger WH, Thomas RB, Sicher RC, George K, Clark JS. 2006. Biomass and toxicity responses of poison ivy (*Toxicodendron radicans*) to elevated atmospheric CO_2. *Proc. Natl. Acad. Sci.* **103**:9086–9089.

ONDCP (Office of National Drug Control Policy). 2001. *The Price of Illicit Drugs: 1981 Through Second Quarter of 2000.* Published by the Office of National Drug Control Policy, Washington, DC, 243 pp.

Patterson DT. 1995. Weeds in a changing climate. *Weed Sci.* **43**:685–701.

Pearson DE, Ortega YK. 2001. An indirect dispersal pathway for spotted knapweed seeds via deer mice and great-horned owls. *Can. Field-Nat.* **115**:354.

Phillips OL, Martinez RV, Arroyo L, Baker TR, Killeen T, Lewis SL, Malhi Y. 2002. Increasing dominance of large lianas in Amazonian forests. *Nature* **418**:770–774.

Rogers CA, Wayne PM, Macklin EA. 2006. Interaction of the onset of spring and elevated atmospheric CO_2 on ragweed (*Ambrosia artemisiifolia* L.) pollen production. *Environ. Health Perspect.* **114**:865–869.

Saghir AR, Markoullis G. 1974. Effects of weed competition and herbicides on yield and quality of potatoes. *Proceedings 12th British Weed Control Conference*. Brighton, UK, pp. 533–539.

Sasek TW, Strain BR. 1990. Implications of atmospheric CO_2 enrichment and climatic change for the geographical distribution of two introduced vines in the USA. *Climate Change* **16**:31–51.

Sharkey TD, Yeh S, Wiberley AE, Falbel TG, Gong D, Fernandez DE. 2005. Evolution of the isoprene biosynthetic pathway in kudzu. *Plant Physiol.* **137**:700–712.

Shea KM, Truckner RT, Weber RW, Peden DB. 2008. Climate change and allergic disease. *J. Allergy Clin. Immunol.* **122**:443–453.

Singer BD, Ziska LH, Frenz DA. 2005. Increasing Amb a 1 content in common ragweed (*Ambrosia artemisiifolia*) pollen as a function of rising atmospheric CO_2 concentration. *Funct. Plant Biol.* **32**:667–670.

Tsugawa H, Tange M, Otsuji J. 1980. Observations on branching and number of leaves in seedlings of Kudzu vines, *Pueraria lobata. Sci. Rep. Fac. Agric. Kobe Univ.* **14**:9–14.

Vanaja M, Jyothi M, Ratnakumar P, Vagheera P, Raghuram Reddy P, Jyothi Lakshmi N, Yadav SK, Maheshwari M, Venkateswarlu B. 2008. Growth and yield responses of castor bean (*Ricinus communis* L.) to two enhanced CO_2 levels. *Plant Soil Environ.* **54**:38–46.

Vengris J. Colby WG, Drake M. 1955. Plant nutrient competition between weeds and corn. *Agron. J.* **47**:213–216.

Wan S, Yuan T, Bowdish S, Wallace L, Russell SD, Luo Y. 2002. Response of an allergenic species, *Ambrosia psilostachya* Asteraceae), to experimental warming and clipping: implications for public health. *Am. J. Bot.* **89**:1843–1846.

Watson WA, Litovitz TL, Klein-Schwartz W, Rodgers GC, Youniss J, Reid N, Rouse WG, Rembert RS, Borys D. 2004. 2003 annual report of the American association of poison control centers toxic exposure surveillance system. *Am J Emerg Med.* **22**:335–404.

Wayne P, Foster S, Connolly J, Bazzaz F, Epstein P. 2002. Production of allergenic pollen by ragweed (*Ambrosia artemisiifolia* L.) is increased in CO_2-enriched atmospheres. *Ann. Allergy Asthma Immunol.* **88**:279–282.

Ye-Ebiyo Y, Pollack RJ, Spielman A. 2000. Enhanced development in nature of larval *Anopheles arabiensis* mosquitoes feeding on maize pollen. *Am. J. Trop. Med. Hyg.* **63**:90–93.

Ziska LH. 2002. Influence of rising atmospheric CO_2 since 1900 on early growth and photosynthetic response of a noxious invasive weed, Canada thistle (*Cirsium arvense*). *Funct. Plant Biol.* **29**:1387–1392.

Ziska LH, Caulfield FA. 2000. Rising carbon dioxide and pollen production of common ragweed, a known allergy-inducing species: implications for public health. *Aust. J. Plant Physiol.* **27**:893–898.

Ziska LH, Emche SD, Johnson EL, George K, Reed DR, Sicher RC. 2005. Alterations in the production and concentration of selected alkaloids as a function of rising atmospheric carbon dioxide and air temperature: implications for ethno-pharmacology. *Global Change Biol.* **11**:1798–1807.

Ziska LH, Epstein PR, Rogers CA. 2008a. Climate change, aerobiology and public health in the Northeast United States. *Mitig. Adapt. Strat. Glob. Change.* **13**:607–613.

Ziska LH, Gebhard DE, Frenz DA, Faulkner S, Singer BD. 2003. Cities as harbingers of climate change: Common ragweed, urbanization, and public health. *J. Allergy Clin. Immunol.* **111**:290–295.

Ziska LH, Morris CF, Goins EF. 2004. Quantitative and qualitative evaluation of selected wheat varieties released since 1903 to increasing atmospheric carbon dioxide: can yield sensitivity to carbon dioxide be a factor in wheat performance? *Global Change Biol.* **10**:1810–1819.

Ziska LH, Panicker S, Wojno HL. 2008b. Recent and projected increases in atmospheric carbon dioxide and the potential impacts on growth and alkaloid production in wild poppy (*Papaver setigerum* DC.) *Clim. Change* **91**:395–403.

Ziska LH, Sicher RC, George K, Mohan JE. 2007. Rising atmospheric carbon dioxide and potential impacts on the growth and toxicity of poison ivy (*Toxidodendron radicans*). *Weed Sci.* **55**:388–292.

8 Weed Management: Herbicides

"Crabgrass can grow on bowling balls in airless rooms, and there is no known way to kill it that does not involve nuclear weapons."

Dave Barry

Introduction

Given the extent of agronomic and environmental damage in managed and unmanaged plant systems by weedy species, and their role in human health, weed management has been, and remains, a primary focus of weed scientists. Practices used to restrict the growth and spread of weeds include prevention, those measures taken to preclude the introduction of a specific weed species, and eradication, the complete removal of any weed species once established. Because eradication is seldom achieved, and because new weeds are being introduced continuously, control remains the purview of many weed scientists. Control involves those management methods that are used to keep a weed species at acceptable levels. Although control methods include the identification and implementation of cultural, mechanical, chemical, and biological options, for most developed countries, including the United States, chemical applications (herbicides) represent the most widely used alternative. Because of the importance of herbicides in weed management, the focus of many weed science textbooks remains herbicide chemistry and management (e.g., Devine et al., 1993; Monaco et al., 2002; Rao, 2000).

In this chapter, we will explore the basic tenets of herbicide usage including application methods, chemistry, and mode of action. We also discuss abiotic and biotic factors that affect the efficacy of herbicides in the field, whether CO_2 and/or climate change are likely to alter these factors, and the resulting implications for chemical weed management. Finally, we review the existing data on climate change and herbicide efficacy to determine if climate change alters efficacy; and if so, whether there is commonality in the observed responses. Our objective is to review the basic principles behind herbicide chemistry, use and application, and to illustrate how climate change is likely to alter the ability of these chemicals to control and manage weeds in the future.

History of Herbicide Use

The decade 1890 to 1900 saw the introduction of sprays for controlling broad-leaved weeds in cereal crops, and the first efforts by the U.S. Army Corps of Engineers, using sodium arsenite, to control aquatic plants in waterways. In 1925, sodium chlorate (directly applied to soil) was first used for killing weeds. The earliest importation (from France) of sodium nitrocresylate, as the first selective weed killer, was in 1934. However, with the advent of World War II, and the shortage of labor for farms, pressure intensified to develop chemical means to reduce weed infestations. By 1945, the introduction of 2,4-D(2,5-dichlorophenoxyacetic acid), a widely used growth regulator that could control broadleaf weeds, ushered in modern weed control. Increasing mechanization and chemical utilization at the farm level, in conjunction with shifts in postwar population demographics, increased herbicide use significantly. By the mid-1960s, for the United States, approximately 55% of farms received herbicides, and by the 2000s, depending on crop, the use of herbicides can exceed 90% (www.nass.usda.gov). This is related, in part, to the development of herbicide-resistant crop cultivars. For example, the introduction and subsequent efficacy of glyphosate (N-(phosponomethyl) glycine), an herbicide with broad-spectrum weed control and minimal environmental impact, has resulted in the development of glyphosate-resistant (Round-up Ready) soybeans, which constitute approximately 90% of the soybeans currently being grown in North America. The success of Round-up Ready soybean has prompted the development of other genetically altered herbicide-resistant crops. At present, acreage treated with herbicides in the United States is approximately twice that of fungicide and insecticide use combined (Figure 8.1).

Application of Herbicides

Application effectiveness depends on timing of herbicide relative to the crop. While there are a number of timing options, the two broad categories are postemergence, where the herbicide is applied directly to emerged plants, or preplant or preemergence, where the herbicide is applied directly to the soil (Fryer and Evans, 1968). However, depending on the stage of growth of the weed relative to the crop, herbicides can be applied preplant, preemergence, at-emergence, postemergent, or some combination thereof. Postemergent applications are made before canopy closure. Herbicides can be applied by spraying directly over the rows, or between rows. They can be incorporated into the soil, or applied to the soil surface. If applied in spray form, the spray can be directed to the tops of leaves, or under the plant canopy. Sprays which include the active chemical ingredient needed to kill the plant are modified by surfactants or wetting agents in order to increase their ability to transfer the active ingredient to the plant surface. Herbicides can be selective or nonselective depending on the mode of action and specific physiology of the plant to be controlled.

Herbicide Characteristics

There are numerous sources that explain in greater depth, the nomenclature, classification, mode of action, and chemistry of herbicides (e.g., Duke, 1985, inter alia). We provide an

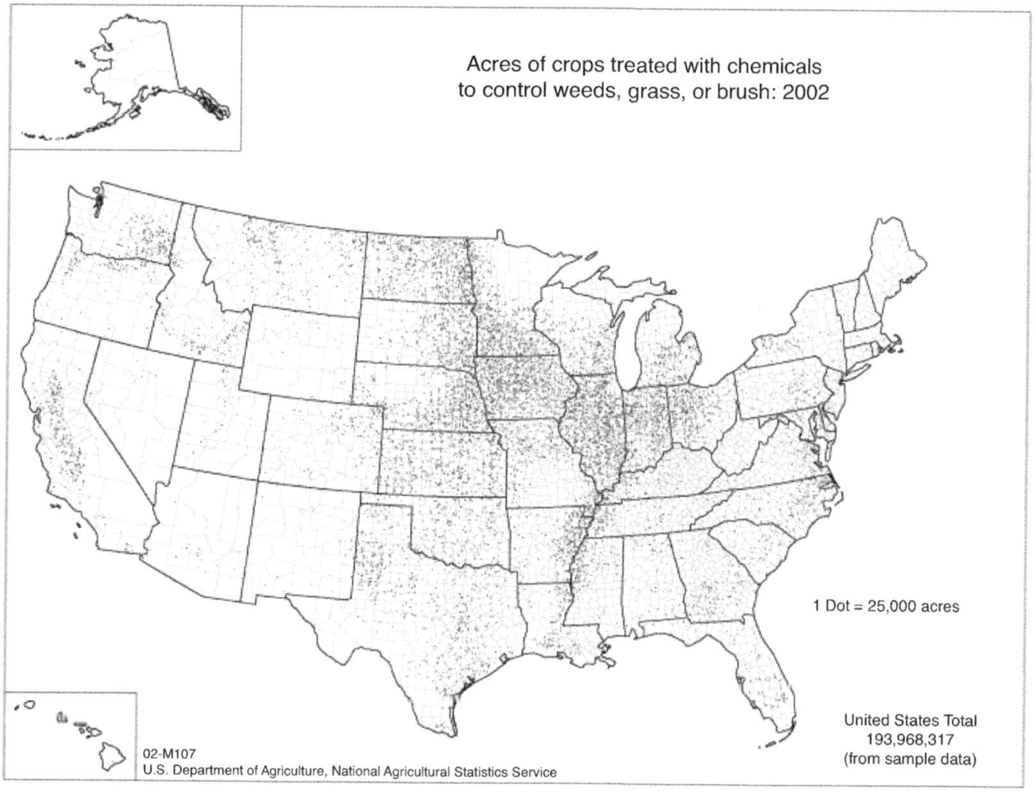

Figure 8.1 Acreage treated with herbicides in 2002. For the same year, 65 million acres were treated for insect control, and 12 million acres for disease control.

abbreviated overview of herbicide characteristics, including mode of action, stressing that the best herbicides are those that are effective in low dosages, have a broad range of uses, are easy to apply and are safe to both the individual and the environment.

Specific information given for each herbicide following certification is reviewed in the *Herbicide Handbook*, a valuable resource for herbicide applicators (WSSA, 2007). Provided information includes nomenclature (common name of the herbicide, chemical family, and manufacturer), chemical and physical properties of the active ingredient (molecular weight, physical, and chemical properties), recommended uses (weeds, dosage, and application times), anticipated plant response (absorption and translocation characteristics, mechanism of action, plant symptomology, and resistance characteristics), longevity in the environment (mobility, volatilization, and soil mobility), toxicological properties (LD_{50}, results of feeding studies, teratogenicity on rats and rabbits, possible effects on wildlife), and specific details on how the chemical was analyzed. Additional sources of information are also provided in the handbook.

All herbicides have three names: the chemical, common, and trade name. There are roughly 200 chemicals classified for herbicide use and the common and chemical names are provided by the WSSA (9th edition handbook). Common names are useful because they do not vary for any parent chemical; consequently, their familiarity among applicators leads to greater consistency of use and increased safety. Conversely, trade names can vary by herbicide, particularly if sold by more than a single company.

For simplicities sake, we will not classify herbicides by chemical similarity, but by primary mechanism of action, using the approach outlined by Monaco et al. (2002) in their herbicide categorization. Additional information on mechanism can be found from Chapters 9–17 in this same reference.

Photosynthetic Inhibitors

These herbicides impede electron transport in the light reactions (Photosystem II) resulting in the production of free radicals, strong biological oxidants that destroy membrane integrity and photosynthetic function. The active ingredient can be translocated from the soil into roots, or applied directly postemergent to leaves. Susceptibility is defined by leaf chlorosis (from bottom to top) followed by necrosis and leaf drop. Herbicides characterized by this mode of action include s-triazines, including the chloro-triazines, atrazine, simazine and cyanazine, methoxy-triazines such as prometon, and methylthio-triazines such as prometryn. This group also includes substituted ureas (e.g., diuron, fluometuron, and linuron) and uracils (e.g., bramcils) as well as benzothiadiazole (e.g., bantazon), benzonitrile (e.g., bromoxymil), and phenylcarbamates (e.g., desmedipham).

Pigment Inhibitors

These herbicides act as inhibitors of the of the carotenoid pigment biosynthetic pathway. Because carotenoids prevent plant pigments from photooxidation, their loss results in chlorophyll destruction and inhibition of photosynthesis. This class of herbicide is usually applied as a preplant or preemergent treatment. Because of the mode of action, symptomology includes a bleached white or translucent leaf appearance. Examples of this type of herbicide include amitrole, clomazone, and fluridone as well as the pyridazinone and isoxazole groups.

Cell Membrane Disruptors

These diverse compounds result in the formation of free radicals that disrupt cell membranes. The site of membrane disruption is classified as B1 (photosystem I disruption), B2 (inhibition of protoporphyrinogen oxidase, the last enzyme of the common branch for chlorophyll and heme biosynthesis), and group C (glutamine synthase, essential for nitrogen metabolism). Examples of B1 include bipyridyliums such as diquat and paraquat, B2 examples include diphenylethers (e.g., lactofen and oxyfluorfen), and group C includes N-phenylphthalimide such as glufosinate. Most herbicides of this type are applied as postemergence; and injury symptoms include leaf desiccation as a result of membrane disruption.

Cytological Growth Inhibitors

These chemicals act to disrupt mitosis in growing root and or shoot tissue. The mitotic disrupters inhibit initial plant cell division and cell wall formation, whereas the root/shoot inhibitors deplete long chain fatty acids from the plasma membranes of plants. Examples of mitotic disruptors include dinitroanilines such as benefin and trifluralin; shoot inhibitors for emerging seedlings include carbamothioates (e.g., EPTC and triallate); seedling root inhibitors include napopamide and siduron, and inhibitors of both shoot and root growth include chloroacetamides such as metolacholor and propachlor and oxyacetamides such as flufenacet. Herbicides of this type are applied as preemergence or preplant incorporated treatments. Symptomology includes thickened, truncated root stems, stems with callus growth, or if shoots do emerge, a twisted tightly rolled appearance.

Inhibitors of Cellulose Biosynthesis

Although diverse in chemical structure, all compounds act to inhibit cellulose biosynthesis with a subsequent collapse of cell structure integrity. Examples include nitriles such as dichlobenil, benzamide, and quinolinecarboxylic acid. Application is usually during preemergence and symptoms include stunted growth and root swelling.

Growth Regulators

The specific physiological mechanism related to growth disruption is unclear; however, many of these compounds act to disrupt or alter hormonal regulation (e.g., auxin activity), nucleic acid metabolism, and protein synthesis. As such, they may result in rapid cell proliferation with twisting and epinasty of the shoot or meristem. Because they work best on meristematic tissue, they may selectively kill broadleaf weeds although grass crops can be injured if applied at the wrong time or at high concentrations. Examples of growth regulators include phenoxy acetic acids such as 2,4-D, phenoxypropionic acids such as dichlorprop, and phenoxy butyric acids such as 2,4-DB. Growth regulators may also include benzoic acids (dicamba) and picolinic acid compounds such as picloram, triclopyr, and clopyralid.

Inhibitors of Lipid Biosynthesis

These chemicals work to inhibit the acetyl coenzyme A carboxylase (ACCase), necessary for lipid biosynthesis in plants. Without fatty acid (lipid) formation, cell integrity is lost, and growth ceases. Because the structure of ACCase is different in broadleaf versus grasses, these chemicals have specific activity against grasses only. Examples include the aryloxyphenoxypropionates (e.g., diclofop and others with the "fop" suffix) and cyclohexanediones (e.g., clefoxydim and others with the "dim" suffix). Most are applied postemergent to unstressed, rapidly growing grassy weeds and can be translocated via the xylem and phloem to growing tissue. Symptomology includes cessation of growth, discoloration of the leaves (usually a reddening), and progressive necrosis beginning at the meristematic tissues and spreading throughout the plant.

Amino Acid Inhibitors

This group of chemicals acts to inhibit amino acid synthesis. They are divided into two groups based on function; those that inhibit the production of aromatic amino acids (phenylalanine, tryptophan, and tyrosine) produced in the shikimic acid pathway by blocking the enzyme 5-enolpyruvylshikimate-3-phosphate synthase (EPSPS); and those that inhibit acetolactate synthase (ALS), an enzyme necessary for the production of branch chain amino acids such as isoleucine, leucine, and valine. These chemicals are nonselective, active only on plant-specific biochemical pathways (i.e., not human or animals; Saari and Mauvais, 1996), and have almost no soil residual times, with low environmental profiles. Because of these characteristics they have become widely used. For example, glyphosate ("Round-up"), an EPSPS inhibitor, is the largest-selling herbicide on the global market (Woodburn, 2000).

The popularity of these herbicides has led to a number of commercial ventures to genetically modify herbicide-resistant crops (e.g., "Round-up Ready" soybean), and this trend is expected to continue. Examples of aromatic amino acid inhibitors include glyphosate and sulfosate; examples of inhibitors of branch chain amino acids include the sulfonylureas (e.g., nicosulfuron, or others with the "furon" suffix), the imidazolinones (e.g., Imzazpyr), the triazolopyrimidines (e.g., diclosulam), and Pyrimidinyloxybenzoates (e.g., Pyrithiobac). Aromatic amino acid inhibitors are applied postemergence to foliage, whereas branch chain amino acid inhibitors can be applied to both foliage and soil. Symptoms include chlorosis and necrosis of all visible plant parts, usually starting at the meristems. Death usually occurs within 4–8 days depending on plant species and concentration used.

Factors that Influence Herbicide Efficacy

The effectiveness of any applied herbicide on weed eradication is influenced by the amount applied, the timing of application relative to the biology of the weed, and the environmental conditions during and after application. Here, we discuss those environmental and biological characteristics likely to influence the uptake and translocation of herbicides and their subsequent efficacy.

Environmental

Environmental (abiotic) factors that can influence herbicide action include soil moisture and temperature, rainfall, wind, relative humidity, air temperature, and light. Soil temperature and moisture directly affect the movement of soil-applied herbicides by affecting herbicide solubility and movement. In addition, soil water availability and temperature may indirectly affect root characteristics such as permeability and herbicide transport via transpirational flow. Rainfall, particularly immediately after application, will affect foliar uptake and translocation especially for water-based herbicides. Wind can interfere directly with surface application and drift; it can also accelerate drying of any solution with a subsequent reduction in chemical absorption. In addition, wind can affect evapotranspiration from the leaf surface or cause cuticular damage, altering soil uptake of herbicides. Conversely, at high temperature and low humidity, wind can accentuate the action of contact herbicides (Muzik, 1976). Relative humidity and air temperature influence evaporation. Increasing evaporation can reduce droplet size and increase herbicide drift. In addition, low humidity accelerates drying of droplets after application. Conversely, high humidity increases leaf retention times and can increase herbicide absorption. Temperature, aside from its role in evaporation, directly affects herbicide volatility. Although rising temperature increases herbicide volatility; increasing temperature can also enhance absorption and translocation of the herbicide, particularly at low humidity (Pallas, 1960). For example, an increase of air temperature from 22 to 32°C resulted in a significant increase in absorption and translocation of melfluidide in soybean (McWhorter and Wills, 1978). Light can also influence herbicide absorption, particularly in conjunction with high temperature as it may increase vaporization and metabolic degradation (Coupland, 1986).

Plant-Based Factors

There are a number of plant-based (biotic) factors that also influence herbicide efficacy. These characteristics are related to both species-specific biology as well as phenotypic environmental responses. Inherent biological characteristics influencing herbicide effectiveness include growth rate, with younger, actively growing plants being, in general, more susceptible to herbicides; branching pattern (e.g., broadleaf versus grasses) which, in turn, influences the amount of surface coverage (and absorption) of postemergent herbicides; leaf surface characteristics, such as cuticular thickness, stomatal number, pubescence, and inherent physiological characteristics, such as the ability to compartmentalize and transport absorbed substances. For example, Rubin and Eshel (1977) found that terbutryn and fluometuron (photosynthetic inhibitors) were selectively translocated depending on species: they accumulated in the root systems of cotton, but translocated to the shoots in the more sensitive snapbean. Differences in uptake and translocation of herbicides reflect different physiological species characteristics and herbicide selectivity.

There are also a number of phenotypic interactions between environmental variables and plant growth and development that affect herbicide efficiency. For example, soil and air temperature affect growth rates of seedlings, and thus the amount of time when plants are most susceptible to herbicides. Temperature and humidity will affect water absorption and transpiration, and hence chemical absorption and movement. In addition, both variables affect leaf development, including cuticular thickness and stomatal number and aperture. Water stress affects herbicide performance by altering foliar uptake. Moisture-stressed plants tend to minimize their leaf surface area and orient leaves perpendicular to the ground to reduce light interception and temperature, with negative consequences for foliar herbicide interception. However, several studies indicate that the impact of moisture stress on herbicide efficacy is dependent on the degree of stress imposed. For example, negative effects of moisture stress on efficacy have been controlled by increasing herbicide rate or alleviating moisture stress within 48 hours of herbicide application (Rossi et al., 1994). Still, weeds that encounter severe water deficits during herbicide application may not respond even if the herbicide rate is increased and moisture stress alleviated. A number of herbicides are dependent on metabolic function; consequently, adverse environmental conditions that reduce physiological performance adversely affect herbicidal mode of action. That is, any environmental variable that reduces photosynthetic activity is likely to alter the action of herbicides that interfere with photosynthesis. For example, if moisture stress results in stomatal closure, this will reduce photosynthesis. When photosynthesis is reduced, other metabolic pathways (e.g., amino acid synthesis) may also slow or stop. If this continues, then chlorophyll demand may also be reduced. Translocation of sugars and carbohydrates may slow. Since a number of herbicides also act to inhibit these processes (e.g., glyphosate and amitrole), their ability to inhibit metabolic function will be negatively affected. Similar photosynthetic patterns may be observed if light or nutrients become limited. Of course, the converse is true as well. That is, if light, humidity, nutrients, temperature, moisture, etc., are optimal for growth, then herbicides that act to inhibit a specific metabolic function will be enhanced.

CO_2, Climate Change, and Herbicide Efficacy: Potential Interactions

Given the direct and indirect effects of rising carbon dioxide and climate uncertainty, several mechanisms are likely to lead to changes in herbicide efficacy and chemical management of weeds. These mechanisms span the biological spectrum from the biochemical to the ecological.

At the biochemical level, rising carbon dioxide and increasing temperatures are likely to increase photosynthesis and alter enzymatic activity and pigment production. Any chemical inhibitors of these metabolic functions (e.g., atrazine) could therefore increase in efficacy. However, protein content per gram of tissue can be reduced with increasing CO_2 concentration (Bowes, 1996). Less protein could, in turn, result in less demand for aromatic and branch chain amino acids, with a possible reduction in the effectiveness of amino acid inhibitors (e.g., glyphosate).

At the leaf level, there may be a number of CO_2- and climate-induced effects that could alter herbicide activity. For example, if rising CO_2 results in greater leaf number or leaf area, this could increase interception of postemergent herbicides. Such an increase could even occur with water stress, as rising CO_2 may increase water use efficiency and subsequent leaf area (Patterson, 1986). Alternatively, increasing CO_2 and/or water stress could result in changes in leaf anatomy and leaf surface characteristics that would reduce herbicide absorption. For example, elevated CO_2 could decrease stomatal number or aperture, whereas prolonged water stress could increase cuticular thickness or leaf pubescence (Ziska and Bunce, 2006). Accumulation of leaf starch, which can occur in C_3 plants exposed to elevated CO_2 (e.g., Arp, 1991), could also interfere with herbicide activity (Patterson, 1995).

At the whole plant level, developmental rates may increase. Although this could increase metabolism, it could also reduce the time weeds spend in the seedling stage, that is, the period of greatest herbicide sensitivity. Allometric changes are also likely, with climate and CO_2 altering allocation to various plant organs, with subsequent effects on translocation. Changes in timing and development of leaves may lead to morphological changes, in leaf shape as well as flowering and pollen production, root:shoot allocation, chlorophyll degradation, etc. (Ziska and Bunce, 2006). Change in root:shoot allocation may be of particular consequence for C_3 perennials that are capable of both sexual and asexual reproduction since increased temperatures and increased metabolic activity could result in greater rhizome and photosynthate supply to below-ground structures than to leaves (Rogers et al., 1992).

At the community or ecosystem level, CO_2 and climate are likely to alter gene flow by altering flowering times or pollen production (e.g., Ziska and Caulfield, 2000); consequently, for crops that have been genetically engineered to be herbicide resistant, gene flow could introduce engineered genes into wild populations (Daniell, 2002; Ellstrand, 2003). Unfortunately, in many instances these wild populations are very similar to the cultivated crop lines (e.g., wild oat and oat, rice and red rice) and pose a significant weed threat; obviously, introduction of genes into these wild populations that confer herbicide resistance is to be avoided.

The potential evolutionary consequences of global change, particularly as it relates to chemical management, have received little attention. Several researchers have documented biotic- or abiotic-induced genetic adaptation (e.g., Lande and Shannon, 1996; Travis and Futyuma, 1993). Researchers have documented adaptation to climate change in *Drosophila subobscura* (Rodriguez-Trelles and Rodriguez, 1998). Chakraborty et al. (2000) has suggested similar changes for pathogens (e.g., *Colletotrichum gloeosporioides*; Chakraborty and Data, 2002). A single study has documented rapid genetic changes in field mustard (*Brassica nigra*) with drought (Franks et al., 2007). However, our understanding of climate-driven evolutionary change, particularly gene flow and hybridization, in agricultural weeds is almost nonexistent (Brasier, 2000). Yet, understanding the extent to which such climatic changes, could spread herbicide resistance is of obvious consequence.

In addition to biological considerations, there are a number of potential abiotic interactions between herbicide application and climate. If extreme weather events happen more often, there may be fewer good times in which to apply herbicides. Similarly, if rain, wind, relative humidity or temperatures (soil and air) are inconsistent, than herbicide application efficacy and retention times will be negatively impacted (Coupland, 1983; Hatzios and Penner, 1982; Muzik, 1976; Patterson et al., 1999). For example, increases in precipitation (in single events or annual means) could increase herbicide leaching and groundwater contamination (Froud-Williams, 1996). Windier conditions could reduce the number of suitable times for application and increase the risk of drift. Similarly, increasing temperatures and drought could increase herbicide volatilization. An economic evaluation based on anticipated climatic change indicated that warming increased pesticide cost variance for corn, potatoes, and wheat, while decreasing it for soybean (Chen and McCarl, 2001). Overall, variation in environmental factors may make application more difficult for a given crop, alter the persistence of the herbicide in the environment, and increase herbicide injury of crops and other nontarget organisms.

CO_2, Climate Change, and Herbicide Management: A Review of Existing Studies

Although there are a number of probable mechanisms by which rising CO_2 and climatic uncertainty could influence herbicide efficacy and chemical management, there are only a handful of studies that have attempted to clarify potential interactions. In a greenhouse experiment, Ziska et al. (1999) examined whether CO_2 affected glyphosate sensitivity of redroot pigweed and common lambsquarters (a C_4 and C_3 species, respectively). Tolerance was determined by measuring growth (biomass) following glyphosate application at 0, 0.112 kg active ingredient (ai) ha^{-1} (10% of manufacturers recommended rate) and 1.12 kg ai ha^{-1} (100%) of recommended rate over a 14-day period in four separate trials that differed in time of application. Although elevated CO_2 had no effect on glyphosate sensitivity in redroot pigweed, the 10% application rate had no effect, and the full, recommended dosage reduced, but did not eliminate, growth in lambsquarters when grown at elevated CO_2 (Ziska et al., 1999). Similarly for a C_3 perennial weed, quackgrass (*Elytrigia repens*), glyphosate tolerance increased for two different cohorts for application rates of 2.24 kg ai ha^{-1} (100% of manufacturers recommended rate for herbaceous perennials) when grown at elevated (770 ppm) CO_2 (Ziska and Teasdale, 2000). Field trials for Canada thistle with both glyphosate and glufosinate (cell membrane disruptor) also showed reduced herbicide efficacy when grown at elevated carbon dioxide (Figure 8.2). Archambault et al. (2001) also reported

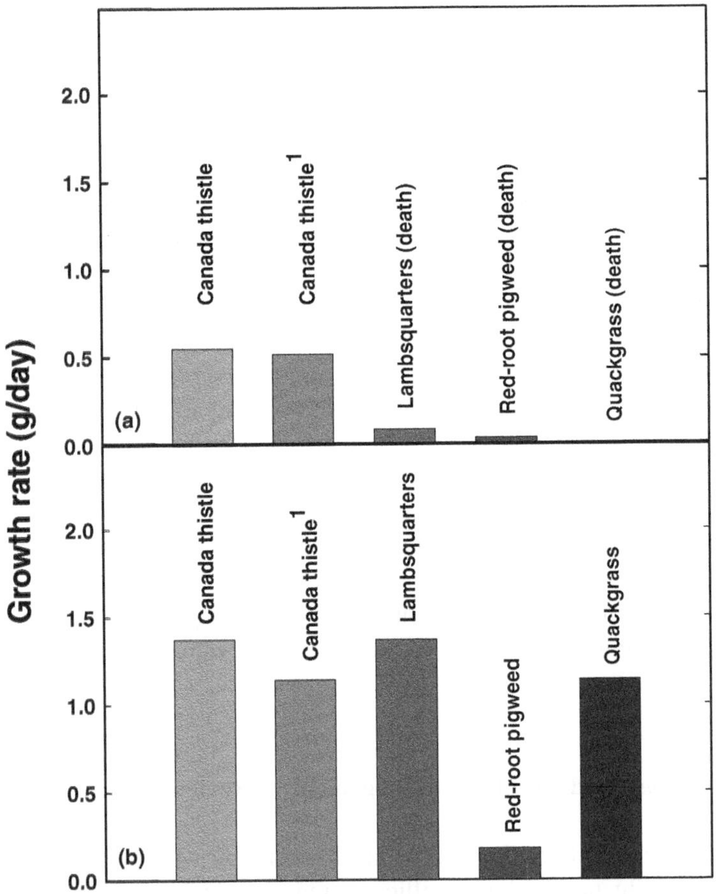

Figure 8.2 Changes in efficacy determined as changes in growth (g day^{-1}) following herbicide application for weeds grown at either current (a) or projected (b) levels of carbon dioxide.

that herbicide efficacy was frequently negatively affected by elevated CO_2, with effects dependent on the mode of action, weed species, and competition. For example, although elevated CO_2 did not affect the response of wild oat to imazemethabenz (an ALS inhibitor), the efficacy of glufosinate-ammonium declined with increased temperatures and/or CO_2 (see Figure 8.1; Archambault, 2007). Archambault (2007) suggested that any economic losses resulting from CO_2 or climate-induced decreases in herbicide efficacy could be compensated by CO_2/climate stimulation of crop yield.

Although it has been suggested (Archambault, 2007) that elevated CO_2 could compensate for the economic cost of increased herbicide use by stimulating seed yield, that suggestion was not supported by an experiment with "Round-up Ready" soybean in the field (Ziska and Goins, 2006). In this experiment, soybean showed a CO_2 by glyphosate interaction, with significant C_3 broadleaf weed biomass remaining viable even after glyphosate application.

Potential Mechanisms: Canada Thistle as a Case Study

Biology

Given the results observed regarding CO_2 and herbicide effectiveness, are there commonalities among possible biological mechanisms related to herbicide efficacy that deserve further exploration? In general, it is acknowledged that active growth is necessary for greater chemical effectiveness: the larger the plant when postemergent herbicide is applied, the less effective the chemical. If CO_2 stimulates plant growth, then reduced efficacy may simply be related to plant size at the time of herbicide application. For both lambsquarters and quackgrass, glyphosate tolerance did increase with plant size; but for any given size when the herbicide was applied, growth at elevated CO_2 resulted in enhanced tolerance to the herbicide relative to the ambient CO_2 condition (Ziska and Teasdale, 2000; Ziska et al., 1999).

What is the basis then for CO_2-induced changes in glyphosate tolerance? Given the diversity of herbicidal modes of action, genetic and phenotypic variability among weeds, and the potential mechanisms through which climate and CO_2 can alter efficacy, it seems unlikely that a common biological explanation is responsible for the range of observed responses. Tolerance could be improved if the absorption of herbicide entering the plant was reduced or if the physiological action of the herbicide was negated in some manner. Short-term CO_2 exposure resulted in stomatal closure and reduced herbicide uptake, and a reduction in stomatal conductance, but had no effect on herbicide tolerance (Ziska and Teasdale, 2000). However, this does not rule out longer term CO_2 responses of leaf characteristics that could affect herbicide absorption.

A simpler explanation for reduced efficacy may be related to CO_2-induced changes in biomass allocation. The longest study on CO_2 and herbicide efficacy to date examined Canada thistle tolerance to herbicides in the field. After each application of glufosinate and glyphosate, above-ground biomass of Canada thistle became completely desiccated

Figure 8.3 Plants in front were sprayed with recommended dose of glyphosate, plants in the back were removed by hand. Chamber on left shows Canada thistle grown at ambient CO_2; plants on the right were grown at ca. 700 ppm CO_2.

and necrotic within 2 weeks (Ziska et al., 2004). At ambient CO_2, herbicide efficacy based on regrowth following application was approximately 90%, but when the herbicide was applied to elevated-CO_2-grown plants, efficacy was only about 10% (see Figure 8.3). Because above-ground biomass was killed regardless of treatment, absorption and mode of action did not appear to be affected by CO_2 treatment. However, elevated CO_2 increased root growth twice as much as shoot growth (see Figure 8.4). This indicated that the differential allocation of additional assimilate generated by greater CO_2 may have preferentially increased root over shoot biomass by the time of herbicide application. This suggests, in turn, that CO_2 stimulation of below-ground biomass could have resulted in a dilution effect of any systemic herbicide (e.g., glyphosate) whereby below-ground structures were not killed. This is consistent with the supposition of Patterson and Flint (1990) who postulated that chemical control could be adversely affected if additional CO_2 resulted in greater root, rhizome, or tuber growth, because each below-ground structure could give rise asexually to a new plant.

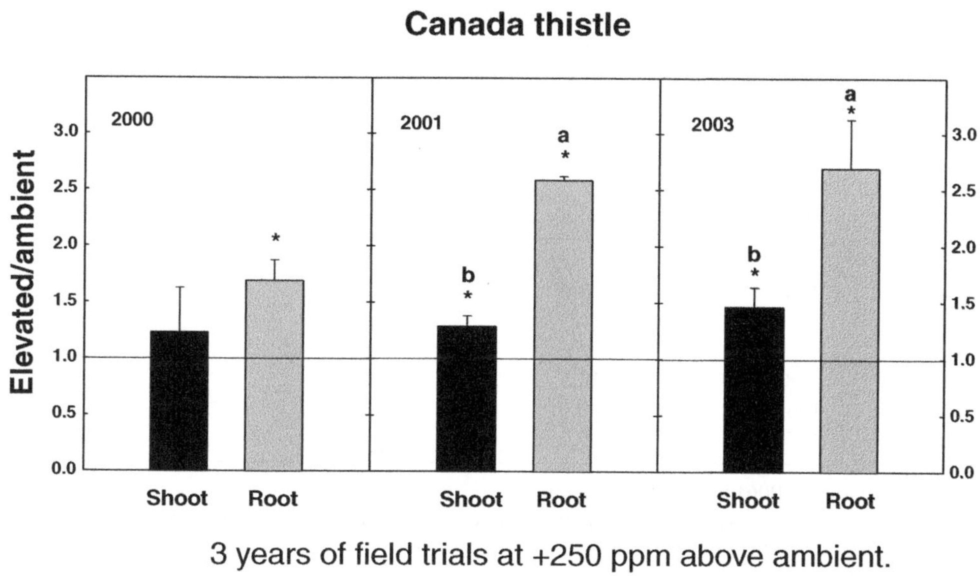

Figure 8.4 We are not clear why invasives would show a strong response to carbon dioxide, there may be a number of mechanistic explanations. However, in the case of Canada thistle, below-ground production (root biomass) appears to be a strong responsive sink when additional carbon dioxide is given. In this experiment, the ratio of elevated to ambient represents the growth of shoots (above ground) and roots (below ground) biomass at current and current +250 ppm CO_2.

Even among pernicious weeds, Canada thistle is well known. According to state noxious weed lists, it was the most troublesome weed in North America in 2000 (Skinner et al., 2000). One possible source of the weed? From hay brought in to feed the horses of General Burgoyne as he invaded the upper Hudson Valley from Canada during the Revolutionary war. Although Burgoyne was eventually defeated at the battle of Saratoga in 1777 (above), the presence of the weed prompted farmers to speculate that the British general had brought it with him from Canada, hence—Canada thistle. (In Australia, the same plant is known as California thistle, since many Australians believe it came from California.) How bad a weed is it? The first record of weed control legislation in the United States was for Canada thistle in 1795 in Vermont, four years after it became a state.

Table 8.1 Averages and the level of statistical significance of the one-way analysis of variance for CO_2 concentration (380 or 630 μmol mol^{-1}) effects on above-ground biomass for weed species associated with field-grown Round-up Ready™ soybean with and without applications of glyphosate (+Gly or –Gly) in 2003 and 2004

	Averages				Level of significance		
Weed type	380	630	+Gly	–Gly	CO_2 effect	Gly effect	Gly × CO_2
2003							
C_4 grasses	138.1	128.3	0.0	266.1		***	
2004							
C_3 broadleaf	48.3	303.9	18.5	333.7	***	***	***
C_4 broadleaf	166.1	257.1	0.0	423.2		***	
C_4 grasses	157.1	23.3	0.5	179.8	(*)	*	

See Ziska and Goins (2006) for additional details.
Data are g/m^2. Note that a significant interaction between CO_2 and glyphosate was observed in 2004 due, in part, to the large growth stimulation of C_3 weeds.
(*) $P < 0.10$; * $P < 0.05$; ** $P < 0.01$; *** $P < 0.001$.

Environmental

Little field-based information is available regarding climatic change and herbicide efficacy. We have previously described how plant communities and the distribution of C_3 and C_4 weeds could change as a function of extreme precipitation (Ziska and Goins, 2006). In that experiment, we removed and mixed the top 20 cm of soil from a fallow agricultural field at the end of each growing season in order to obtain uniform distribution of the weed seed bank. We then exposed a portion of that soil each year to elevated (250 ppm above ambient) CO_2 to determine possible changes in weed species associated with field-grown Round-up Ready soybean with and without applications of glyphosate (Ziska and Goins, 2006). Unexpectedly, precipitation shifted dramatically between years, with a record year in 2003 (highest since 1895), followed by a near normal year in 2004. Interestingly, precipitation itself was a strong selection factor for weed populations, with only shallow rooted weedy grasses able to grow in the anoxic environment created by the record wet year. Most of these weedy grasses were C_4, and elevated CO_2 did not affect biomass or physiology, so no interaction was observed between CO_2 and glyphosate (see Table 8.1). However, when precipitation was near normal levels the following year, a range of different weed species including C_4 broadleaf species and grasses as well as C_3 broadleaf weeds germinated and grew. Increased CO_2 affected broadleaf weeds, stimulating their growth, and making herbicidal (glyphosate) control more difficult.

This study, while preliminary, suggested that the environmental factors that influence the composition and rate of growth of the weed community will also indirectly impact chemical efficacy. For example, chemical efficacy was unaffected by CO_2 in a rainy year that favored the growth of C_4 grasses. If, in fact, CO_2 effects on herbicide tolerance are associated more with C_3 species (e.g., Ziska et al., 1999), then any environmental factor that alters the ratio of C_3 to C_4 weeds (e.g., light and nitrogen deposition) could result in differential herbicide effects.

Should We Worry?

Given the remarkable effectiveness of herbicides, is it necessary to be concerned about any potential exacerbation of weed damage from climate change? After all, many of the most widely used herbicides are effective at low dosages, easy to apply, and, in a number of cases, are applied to crops that have been genetically engineered to be tolerant of the herbicide. One can argue therefore, that even if weeds are more of a nuisance in a warmer, wetter (or drier), higher CO_2 world, so what?

There are a number of implicit assumptions in this argument. The first assumption is that climate and or CO_2 will be independent of any change in chemical efficacy. While there is a great deal of uncertainty regarding the mechanism, a number of studies show that climate and/or CO_2 can influence (in many cases, negatively) the overall efficacy of the chemical applied. Although the mechanism(s) associated with the change in efficacy are not completely understood, there are a number of physiological and environmental factors associated with climate change that are likely to reduce our ability to control weeds by chemical means. It can be argued that CO_2 or climate-induced changes in chemical efficacy are irrelevant given the rate of atmospheric CO_2 increase (i.e., other, better, herbicides will be developed in the future). However, we would stress that herbicide use can persist for decades. For example, 2,4-D has been in widespread use since 1950, coinciding with an approximate 20% increase in atmospheric CO_2 (315–384 ppm). Also, at present, a number of commercial ventures have a financial interest in developing genetically modified crops specifically tolerant to a given herbicide. Consequently, the use of these associated herbicides (e.g., glyphosate) could continue over long periods.

Or, it may not. The second assumption is that the efficacy of these herbicides will continue in the future as in the past. However, the ease and simplicity of herbicides such as glyphosate carries a long-term cost. That is, greater usage over larger areas increases selection pressure and increases weed resistance to the herbicide. Selection pressure and altered populations are, of course, the cornerstone of evolution; familiar to anyone in the medical community who has encountered antibiotic resistance. How quickly is such resistance spreading in weed species? To get a sense of this, the weed science society of America keeps track of herbicide-resistant weedy biotypes, by weed, by mode of action of the herbicide, and by other categories (www.wssa.gov). Perhaps most illuminating is the current global distribution of weeds that are resistant to one or more herbicides, shown in Figure 8.5.

One final assumption is that any increase in herbicide application as a management tool to control additional weed impacts related to CO_2 and/or climate may ignore long-term environmental costs. We would argue that increased use of herbicides, or any pesticide, to compensate for reduced efficacy will carry a public health risk (see Chapter 7). Such an environmental risk is likely to be exacerbated if, in fact, the easy to use, low-toxicity herbicides are indeed the first to develop resistance in the weed population, since this could increase reliance on less-used, more toxic chemicals. There is also an economic cost associated with changes in the frequency of application or increased concentration of the herbicide. Indeed, from an economic perspective, any increased profits associated with elevated CO_2 and/or temperature could be negated by additional costs of weed control (Ziska and Goins, 2006).

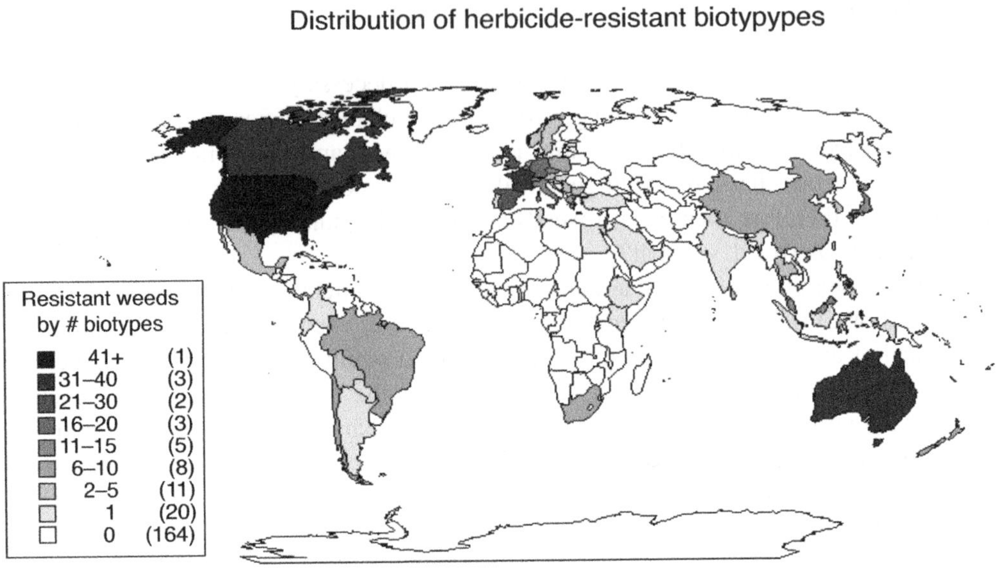

Figure 8.5 From weedscience.com, the distribution of herbicide resistance in weedy biotypes by country. The overall number of weed resistance biotypes has increased 6× since 1980.

Georgia as a Test Case

For much of the southeastern United States, extreme drought was observed from 2005 to 2007. In addition, for Georgia, the incidence of glyphosate-resistant Palmer amaranth increased markedly from one county in 2004 to ten counties as of 2007 (Culpepper et al., 2008a). Palmer amaranth is a significant or troublesome weed in cotton, with two plants per 6 m of row reducing yield by up to 23% (Culpepper et al., 2008a). Can the response of growers to the combination of climatic extremes and increasing herbicide resistance provide some insight as to how farmers could adapt chemical management to meet these new challenges?

The challenges have been met, in part, by implementing what has been termed, "Herbicide Resistance Management." This includes such practices as slowing the spread of glyphosate resistance by rotating with conventional cotton cultivars (i.e., no glyphosate application), increased crop rotation, use of herbicides with different mode of action, less reliance on preplant herbicides (because little moisture has been available to move the herbicide into the soil), and increased use of other, nonchemical means to control weeds, including mechanical tillage and hand-pulling (Culpepper et al., 2008b).

Future Management

As there is no one means by which CO_2 and climate will alter herbicide efficacy, there is probably no one strategy for chemically managing weeds in the future. If CO_2 and/or temperature shorten the seedling stage (i.e., time of greatest herbicide sensitivity), then earlier spraying may be an option; however, earlier spraying could also result in weed regrowth prior to canopy closure. Alternatively, increasing efficacy of surfactants, or other chemical carriers of the herbicide, could increase the ability of the active ingredient to be absorbed even under high wind or rainfall conditions (Froud-Williams, 1996). One promising adaptation strategy may be to build upon existing environmental/herbicide performance profiles such as those described by Coupland and Caseley (1981) and Coupland (1983) for quackgrass and glyphosate, but to include CO_2 and temperature profiles for herbicide efficacy as a function of crop–weed associations (Archambault et al., 2001).

Summary

Given the economic and environmental harm caused by weeds, one of the fundamental objectives of weed biologists is to manage destructive weed populations in order to minimize their impact. In developed countries, chemical application remains the principal means of management. In this chapter, therefore, we have focused on the history of herbicides, their methods of application, their characteristics, particularly their physiological mode of action, and the biotic and abiotic factors that are known to influence their efficacy. In addition, we have examined, both from a theoretical and practical perspective, how CO_2 and climate affect that efficacy. For those data that suggest a reduction in effectiveness, primarily with rising CO_2 levels, we have offered several (as yet untested) possible mechanisms. Looking forward, chemical control practices may need to change to cope with emerging challenges related to increasing herbicide resistance and extreme variation in climate parameters (e.g., drought). Adaptation of herbicide management to these challenges deserves more time and attention than it has received. In the following chapter, we will continue our examination of weed management in the context of climate change, focusing on nonchemical control of weeds, including cultural, mechanical, biological, and integrated forms of control.

References

Archambault DJ. 2007. Efficacy of herbicides under elevated temperature and CO_2. In: Newton PCD, Carran A, Edwards GR, Niklaus PA (eds) *Agroecosystems in a Changing Climate*. CRC Press, Boston, MA, pp. 262–279.

Archambault DJ, Li X, Robinson D, O'Donovan JR, Klein KK. 2001. *The Effects of Elevated CO_2 and Temperature on Herbicide Efficacy and Weed/Crop Competition*. Report to the Prairie Adaptation Res. Coll. No. 29.

Arp WJ. 1991. Effects of source–sink relations on photosynthetic acclimation to elevated CO_2. *Plant, Cell Environ.* **14**:869–875.

Bowes G. 1996. Photosynthetic responses to changing atmospheric carbon dioxide concentration. In: Baker NR (ed.) *Photosynthesis and the Environment*. Kluwer Publishing, Dordrecht, the Netherlands, pp. 387–407.

Brasier CM. 2000. The rise of the hybrid fungi. Nature **405**:134–135.

Chakraborty S, Data S. 2002. Polycyclic infection by *Colletotrichum gloeosporioides* at high CO_2 selects for increased aggressiveness [Abstract]. *Phyopathology* **92**:S13.

Chakraborty S, Pangga IB, Lupton J, Hart L, Room PM, Yates D. 2000. Production and dispersal of *Colletotrichum gloeosporioides* spores on *Stylosanthes scabra* under elevated CO_2. *Environ. Pollut.* **108**:381–387.

Chen CC., McCarl BA. 2001. An investigation of the relationship between pesticide usage and climate change. *Clim. Change* **50**:475–487.

Coupland D. 1983. Influence of light, temperature and humidity on the translocation and activity of glyphosate in *Elymus repens*. *Weed Res.* **23**:347–355.

Coupland D. 1986. The effect of environmental factors on the performance of fluazifop-butyl against *Elymus repens*. *Ann. Appl. Biol.* **108**:353–363.

Coupland D, Caseley JC. 1981. Environmental influences on the effects of glyphosate on *Agropyron repens*. *Proc. Conf. Grass Weeds Cereals UK*, pp. 177–186.

Culpepper AS, Whitaker JR, MacRae AW, York AC. 2008a. Distribution of glyphosate resistant Palmer amaranth (*Amaranthus palmeri*) in Georgia and North Carolina during 2005 and 2006. *J. Cotton Sci.* **12**:306–310.

Culpepper AS, York AC, Kichler J. 2008b. University of Georgia herbicide programs for controlling glyphosate-resistant Palmer amaranth in 2008 cotton. *Circular*, No. 924. Cooperation Service of the State of Georgia, Tifton, GA.

Daniell H. 2002. Molecular strategies for gene containment in transgenic crops. *Natl. Biotechol.* **20**: 581–586.

Devine M, Duke SO, Fedtke C. 1993. *Physiology of Herbicide Action*. Prentice Hall, Englewood Cliffs, NJ, 441 pp.

Duke SO. 1985. *Weed Physiology: Vol. 2, Herbicide Physiology*. CRC Press, Boca Raton FL, 243 pp.

Ellstrand NC. 2003. Current knowledge of gene flow in plants: implications for transgene flow. *Philos. Trans. R. Soc. Lond. B* **358**:1163–1170.

Franks SJ, Sim S, Weis AE. 2007. Rapid evolution of flowering time by an annual plant in response to a climate fluctuation. *Proc. Natl. Acad. Sci. U. S. A.* **104**:1278–1282.

Froud-Williams RJ. 1996. Weeds and climate change: implications for their ecology and control. *Asp. Appl. Biol.* **45**:187–196.

Fryer JD, Evans SA. 1968. *Weed Control Handbook*. Blackwell Press, Oxford, UK, 402 pp.

Hatzios KK, Penner D. 1982. Influence of environmental factors on herbicide by higher plants. In: Hatzios KK, Penner D (eds) *Metabolism of Herbicides in Higher Plants*. Burgess Publishing Company, Minneapolis, MN, pp. 93–100.

Lande R, Shannon S. 1996. The role of genetic variation in adaptation and population persistence in a changing environment. *Evolution* **50**:434–437.

McWhorter CG, Wills GD. 1978. Factors affecting the translocation of ^{14}C-mefluidide in soybeans (*Glycine max*), common cocklebur (*Xanthium pennsylvanicum*) and johnsongrass (*Sorghum halepense*). *Weed Sci.* **26**:382–388.

Monaco, TJ, Weller SC, Ashton FM. 2002. *Weed Science, Principles and Practices*, 4th edn. John Wiley and Sons, Hoboken, NJ, 700 pp.

Muzik TJ. 1976. Influence of environmental factors on toxicity to plants. In: Audus LJ (ed.) *Herbicides: Physiology, Biochemistry, Ecology*. Academic Press, New York, pp. 203–247.

Pallas JE Jr. 1960. Effects of temperature and humidity on foliar absorption & translocation of 2,4-dichlorophenoxyacetic acid & benzoic acid. *Plant Physiol.* **35**:575–580.

Patterson DT. 1986. Responses of soybean (*Glycine max*) and three C_4 grass weeds to CO_2 enrichment during drought. *Weed Sci.* **34**:203–210.

Patterson DT. 1995. Weeds in a changing climate. *Weed Sci.* **43**:685–701.

Patterson DT, Flint EP. 1990. Implications of increasing carbon dioxide and climate change for plant communities and competition in natural and managed ecosystems. In: Kimball BA, Rosenburg NJ, Allen LH, Jr. (eds.), *Impact of Carbon Dioxide, Trace Gases and Climate Change on Global Agriculture*, ASA Special Publication No. 53, American Society of Agronomy, Madison, WI, pp. 83–110.

Patterson DT, Westbrook JK, Joyce RJV, Rogasik J. 1999. Weeds, insects and diseases. *Clim. Change* **43**:711–727.

Rao VSP. 2000. *Principles of Weed Science*. Oxford and IBH Publishing, New Delhi, Calcutta, 568 pp.

Rodriguez-Trelles F, Rodriguez MA. 1998. Rapid micro-evolution and loss of chromosomal diversity in *Drosophila* in response to climate warming. *Evol. Ecol.* **12**:829–838.
Rogers HH, Peterson CM, McCrimmon JN, Cure JD. 1992. Response of plant roots to elevated atmospheric carbon dioxide. *Plant Cell Environ.* **15**:749–752.
Rossi FS, Neal JC, Semesac AF. 1994. Alleviating the antagonistic effect of moisture stress on smooth crabgrass (*Digitaria ischaemum*) control with fenoxaprop. *Weed Sci.* **42**:418–423.
Rubin B, Eshel Y. 1977. Absorption and translocation of terbutryn and fluometuron in cotton (*Gossypium hirsutum*) and snapbeans (*Phaseolus vulgaris*). *Weed Sci.* **25**:499–505.
Saari LL, Mauvais CJ. 1996. Sulfonylurea herbicide-resistant crops. In: Duke SO (ed.) *Herbicide Resistant Crops*. CRC Press, Boca Raton, FL, pp. 127–143.
Skinner K, Smith L, Rice P. 2000. Using noxious weed lists to prioritize targets for developing weed management strategies. Weed Sci. **48**:640–644.
Travis J, Futyuma DJ. 1993. Global change: lessons from and for evolutionary biology. In: Kareiva PM, Kingsolver JG, Huey RB (eds) *Biotic Interactions and Global Change*. Sinauer, Sunderland, MA, pp. 251–263.
Weed Science Society of America. 2007. *Herbicide Handbook*, 9th edn. Weed Science Society of America, Champaign, IL, 240 pp.
Woodburn A. 2000. Glyphosate: production, pricing and use worldwide. *Pest Manage. Sci.* **56**:309–312.
Ziska LH, Bunce JA. 2006. Plant responses to rising carbon dioxide. In: Morison JIL, Morecroft MD (eds) *Plant Growth and Climate Change*. Blackwell Publishing, Oxford, UK, pp. 17–47.
Ziska LH, Caulfield FA. 2000. Rising carbon dioxide and pollen production of common ragweed, a known allergy-inducing species: implications for public health. *Aust. J. Plant Physiol.* **27**:893–898.
Ziska LH, Faulkner SS, Lydon J. 2004. Changes in biomass and root:shoot ratio of field-grown Canada thistle (*Cirsium arvense*), a noxious, invasive weed, with elevated CO_2: implications for control with glyphosate. *Weed Sci.* **52**:584–588.
Ziska LH, Goins EW. 2006. Elevated atmospheric carbon dioxide and weed populations in glyphosate treated soybean. *Crop Sci.* **46**:1354–1359.
Ziska LH, Teasdale JR. 2000. Sustained growth and increased tolerance to glyphosate observed in a C_3 perennial weed, quackgrass (*Elytrigia repens* (L.) Nevski), grown at elevated carbon dioxide. *Aust. J. Plant Physiol.* **27**:159–164.
Ziska LH, Teasdale JR, Bunce JA. 1999. Future atmospheric carbondioxide concentrations may increase tolerance to glyphosate. *Weed Sci.* **47**:608–615.

9 Weed Management: The Rest of the Story

"Free Weeds . . . [You] Pick 'Em"

Anonymous

Introduction

Herbicides have, from the land manager's perspective, a number of desired traits, including ease of use, effectiveness, and cost. However, as pointed out by Ross and Lembi (1999), the widespread availability and use of chemical control has led to the mistaken impression that effective containment of weeds can always be achieved *after* the weed has been introduced. This idea is synergistic with another concept, that of "economic threshold"; that is, weeds are fine until they begin to impact the economics of the land in question. Yet, there are a number of other, nonchemical weed management strategies, that are designed to prevent the initial introduction of the weed, identify the weed once it does occur, and cultural, mechanical, and biological methods to manage, or if possible, eradicate, the weed. As herbicide resistance becomes more prevalent, these alternative strategies, either singly or in combination, may become more efficacious with respect to long-term weed management.

In this chapter, we will examine these weed management strategies including a more in-depth analysis of prevention, identification, and nonchemical control (cultural, physical, and biological) as well as a discussion of integrated weed management. While we will present a traditional analysis of such practices, we will also discuss those physical and biological features associated with CO_2 and climatic change, examine how they could affect nonchemical weed control, and provide suggestions for how these impacts could be incorporated into future research efforts.

Prevention

The purpose of preventive management is to stop the introduction of a given weed species (via seed or vegetative means) into any area where it has not previously occurred. In many instances prevention is not imposed by scientific application but by legal means. For example, quarantines are imposed to isolate and limit the spread of a number of noxious or invasive

Weed Biology and Climate Change, Lewis H. Ziska and Jeffrey S. Dukes, © 2011 Blackwell Publishing Ltd.

weeds, weeds that are recognized as having a significant impact on plant communities even at low populations. Laws associated with quarantines are applied to all systems associated with transport whereby such seed could be spread, including soil, mulches, produce, and farm equipment. These laws may extend to seed certification to ensure seed purity for crops, or lack of weed seed for hay and forages. There may be laws that require eradication of certain weeds, where they appear on land that is privately or publicly owned.

At the national level, the Federal Noxious Weed Act (passed in 1975) was designed to prevent weeds from becoming established in the United States and has allowed USDA inspectors to turn back imports that harbor noxious weeds. The Federal Interagency Committee for the Management of Noxious and Exotic weeds (FICMNEW, www.fs.fed.us/ficmnew/index.shtml), a formal partnership between 16 federal agencies, was designed to coordinate regulatory and control efforts to manage invasive plants across the United States by developing and sharing scientific and technical information.

Unfortunately, there is not always agreement among policymakers at the State or Federal level as to what constitutes a significant weed threat. The USDA's Natural Resource Conservation Service (NRCS) maintains the PLANTS database (www.plants.usda.gov) which is a primary resource for helping private and public landowners in detecting invasive weeds. A search of the plant's database reveals that field bindweed, while considered a noxious weed in 22 states, is not considered a federally noxious weed, nor is purple loosestrife; whereas Melaleuca, *Melaleuca quinquenervia*, is considered a noxious weed by eight states and the federal government. While this piecemeal approach reflects isolated species introductions, it seems inadequate to reflect the rapidly changing nature of global trade; increasing globalization intensifies the need for a coordinated response among federal and state agencies. In addition, increasing trade should be viewed in terms of changing vulnerabilities; weeds that heretofore had not been suitable for certain areas based on their former climates may now (or soon) pose a different threat level when introduced. For example, inferred climatic tolerances of invasive weeds in the western United States suggest areas of potential range expansion and contraction under climate change (Bradley et al., 2009).

Detection

Rapid detection of a given weed may not be necessary if a weed establishes slowly and persists indefinitely at low populations. Most growers and land managers recognize that some weed species are common, but do not pose an immediate threat. However, with increasing recognition that invasive or noxious weeds may pose a significant risk even at low populations, and awareness that the number of herbicide-resistant weeds is increasing, the idea that weeds do not pose a serious danger until they reach a given economic threshold is being questioned. Given the fecundity of many weed species, a single plant is capable of establishing a large population in a short period. If, for example, a few herbicide-resistant Palmer amaranth plants are not detected (see previous chapter), this may result in a significant increase in weed management costs. Therefore, information regarding spatial and temporal changes in such weed populations can be essential for early detection and rapid response (EDRR).

The EDRR concept has become a fundamental practice of the USDA's Animal Plant Health Inspection Service (APHIS), NRCS, and US Forest Service efforts to detect and respond to invasive species (including weeds, pathogens, and insect pests). Much of the early detection for agriculture occurs at the state or county level with extension programs and/or staff in the state departments of agriculture or natural resources who are responsible for identifying and reporting the appearance of invasive species. For example, the first detection of Asian soybean rust in the United States occurred on an LSU Agricultural Center Research Farm in the southern part of East Baton Rouge Parish. Within days of confirmation the Louisiana Department of Agriculture and Forestry, and APHIS plant protection and quarantine teams had sampled soybean and kudzu plants in a 100-mile radius of the original discovery (see USDA Public Soybean Rust Web site, www.sbrusa.net). Mode of entry into the United States was uncertain; however, the distribution pattern of the disease in South Louisiana and Mississippi, and the interval between Hurricane Ivan and detection, suggested that the spores may have been transported from South America by this storm. In addition to any role climate may have played in transport, additional interest has been generated in detecting and mapping kudzu populations, since kudzu is a carrier of the disease. In this context, climate change is a consideration since kudzu establishment is regulated, in part, by thermal minimum temperatures (Sasek and Strain, 1990), and the recent change in plant hardiness zones associated with global warming suggests a northward migration of kudzu (Figure 9.1).

Efforts to combat invasive plant species (IPS) in natural areas are being enhanced by a number of empirical models (for a partial listing of available models, go to http://invasivespecies.nbii.gov/models.html). Briefly, these models fall along three broad categories: (a) Models that estimate a species' climatic tolerances based on its geographical range native and introduced habitats and identify additional areas with potentially suitable climates (e.g., CLIMEX; Sutherst et al., 1999); (b) evaluation of common biological traits among organisms known to be invasive; and (c) a "risk-assessment" approach that evaluates intrinsic and extrinsic factors related to potential introduction and success of a given species (e.g., Rejmanek, 2000).

Detection efforts are also being aided by a number of scientific tools. For example, Gerhards and Christensen have employed real-time weed detection in sugarbeet, maize, and winter wheat using computer analysis of global positioning system data and digital images from cameras mounted on sprayers in the field (Gerhards and Christensen, 2003). Remote sensing offers an additional tool to detect the occurrence of invasive species (e.g., Lawrence et al., 2005; Parker-Williams and Hunt, 2004), particularly invasive plants. However, there is some concern regarding whether such techniques are sufficiently sensitive to detect weeds at low population densities (i.e., detection may not occur until a large population is already established). There is currently an effort underway to link predictive temporal geospatial models with remote sensing to improve performance in regard to climate change and the appearance of invasive species (e.g., Graham et al., 2008).

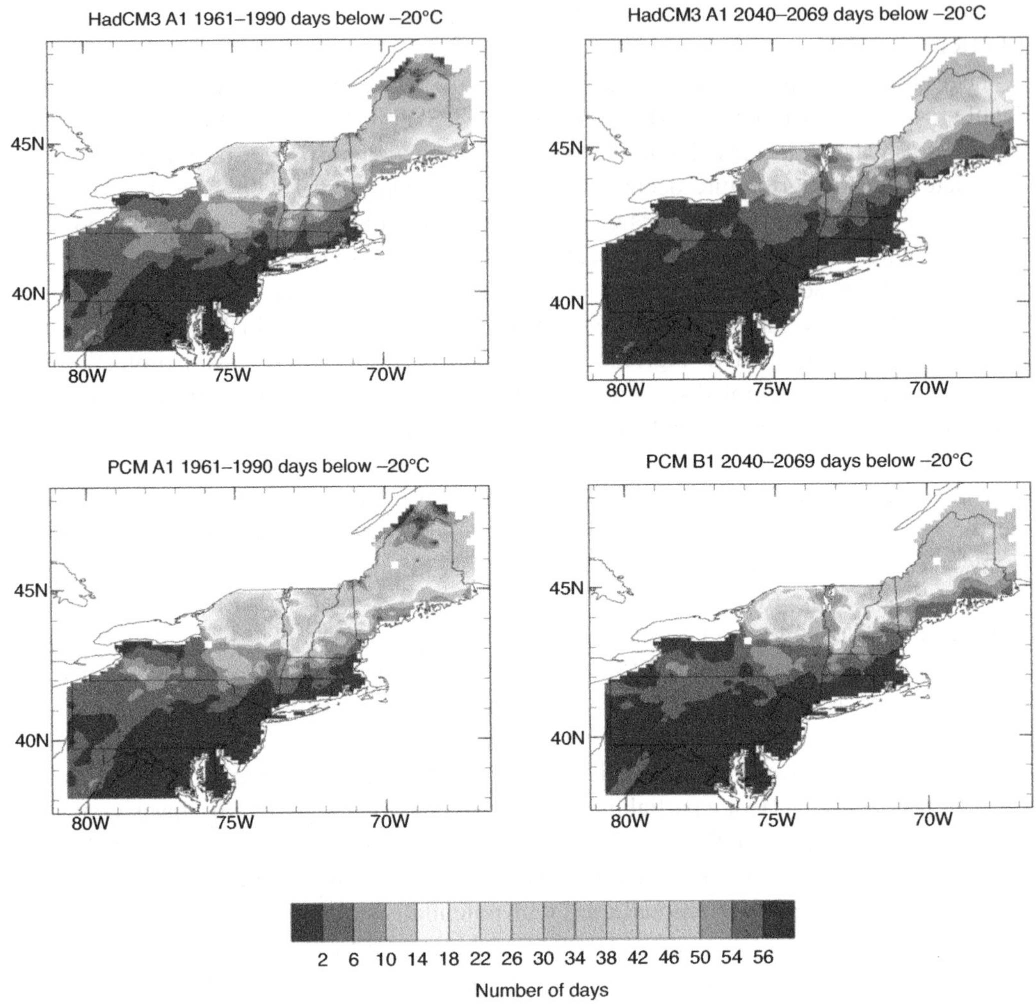

Figure 9.1 Current and projected change in the number of days below −20°C has been shown to be a key factor in the establishment and persistence of kudzu, an invasive perennial weed that is also a host of Asian soybean ruts. (Data courtesy of Katharine Hayhoe, Texas Tech University.)

Physical Control

By Hand

For the large majority of global farmers (250 out of 350 million), hand pulling or hoeing remain the primary methods of weed control (Hill, 1982; Melander et al., 2005; Figure 9.2). Weeding by hand accounts for up to 60% of the total preharvest labor input in the developing world (Webb and Conroy, 1995). Hand pulling may be particularly efficacious when a known noxious weed first appears, or if the area is small or if the crop is of high

Figure 9.2 Hand weeding of Kharif (wet season) rice in Uttarakhand, Indo-Gangetic Plain. (Photo courtesy of Dr. David Johnson, International Rice Research Institute.)

value. Small populations of weeds that are known to be, or are becoming herbicide, resistant should also be removed by hand or hoe. However, depending on farm size, and the degree of infestation, hand pulling may not be a viable option because of time and/or cost. In addition, perennial weeds, which may regrow from below-ground structures, will require multiple visits.

Mowing

Mowing can be an effective means of weed control for perennials if done repeatedly. It is useful for small spaces (such as your front lawn), provided the growing points of the plants in question are above the cutting surface. Those plants with meristems below the cutting surface (e.g., plantain, dandelion, most grasses) will not be affected. Warmer climates, more CO_2, and greater precipitation could, potentially, increase overall plant growth and increase mowing frequency.

Tillage

If farm mechanization is available, tillage is almost certain to occur. Tillage is a form of mechanical control by tractor cultivation using a given implement to manage weed populations. Tilling a field controls weeds by burying seedlings, physically tearing shoots from roots, stimulating germination inducing seed desiccation, or by forcing perennial weeds to utilize stored reserves. Tillage may also break up and aerate soil, prepare seed beds and thin crops to a desired density. Tillage can vary from primary (initial, usually deep, break-up of soil at the beginning of the growing season) to secondary tillage (usually lighter passes less than 10 cm). The type of implement (e.g., disc plow, harrows, sweeps, knives, and brush-rakes) depends on a number of factors including soil type, crop, acreage, and the degree of mechanization (animals or engines). Tillage may be particularly effective in controlling weeds between rows, less so for perennial weeds, or weeds within rows, particularly if the weed resembles the crop.

Different types of tillage exist, usually classified by the degree to which tillage is used. For example, there is a distinction between conventional, minimum, and no-till systems. Reductions in tillage help to reduce energy and labor costs and to conserve soil. However, as tillage is decreased, greater use of herbicides may be necessary in order to achieve satisfactory weed control. A more comprehensive and detailed review of tillage and mechanical weed control is given in texts by Ross and Lembi (1985), Zimdahl (1993), and Vincent et al. (2001).

There are several means by which climate and CO_2 could affect tillage. Crop plants may be stimulated by greater CO_2, resulting in a faster time to vegetative cover, but less time for mechanical tillage. Precipitation extremes of drought and flood can also limit field operations. Climate or CO_2 induced increases in seed production and germination discussed in Chapter 1 may also result in enhanced tillering. Finally, it has been suggested that rising CO_2 levels could result in further belowground carbon sequestration, particularly in roots, or in rhizome growth of perennial weeds, with subsequent increases in asexual reproduction (e.g., Rogers et al., 1994). Data for invasive weeds are consistent with this hypothesis, indicating a significant increase in root to shoot ratio for four of five invasive species with twentieth century increases in atmospheric carbon dioxide (Ziska, 2003). If greater belowground growth for these weeds results in greater asexual reproduction, this would have a negative effect on tillage since this type of control would increase weed spread. For example, as noted in the previous chapter, Canada thistle allocates greater carbon from rising CO_2 to roots over shoots. Disking such a weed grown with elevated CO_2 would result in the rending of the overall root system and a significant increase in the number of potential Canada thistle seedlings.

Fire

Equipment can be attached to tractors or other vehicles that can be used to flame or heat weeds in managed crop systems as a means of control. The control is particularly effective if the crop has a thick or woody stem and is less susceptible than the herbaceous weed (e.g., in cotton management). Flaming has a greater degree of effectiveness for plants between rows, or in border areas (e.g., near irrigation ditches). However, as energy costs increase, or during droughts, flaming may either be economically or environmentally prohibitive.

Chaining

The use of a heavy chain dragged between two tractors is an effective means to clear brush and small trees and has been particularly effective in rangeland management as a means to promote the growth of annual grasses. Chaining is usually most successful if soils are light and the topography is relatively level.

Interestingly, evidence from Morgan et al. (2007) supports an earlier hypothesis (Polley et al., 1997) that incursion of woody plants into grasslands around the world may be associated with the increase in atmospheric carbon dioxide and the selection of C_3 shrubs over C_4 grasses. Encroachment of shrubs and other woody perennials into grasslands has consequences for the efficacy of chaining and other methods of vegetation management in rangelands.

Cultural Control (Farm Management)

Flooding

Although flooding can be used as a means of physical control, primarily by land managers who wish to eliminate a stand of weeds by maintaining flooded conditions, it is, on a global basis, one of the most effective nonchemical means to control weeds culturally in rice (paddy) systems. Rice is the staple food for much of the global population, particularly in poor countries. Weeds, in turn, are recognized as the major biological constraint for rice production worldwide (Rosenzweig and Hillel, 1998). Flooding can improve soil chemical conditions and imposition of an anaerobic environment prevents the establishment of many weed species. Because of the importance of rice in food systems, and the impact of weeds on rice production, the imposition of flooded conditions is one factor that allows paddy or irrigated rice to achieve the highest yields among rice systems (e.g., upland, rainfed; Watson, 2002).

Because of the importance of water in rice production and weed control, declining water for irrigation is likely to have significant effects not only on paddy rice production, but also on how rice is grown. In Asia, for example, competition for fresh water between cities and rural areas has increased the cost of water delivery. Rivers such as the Indus, Ganges, and Brahmaputra have supplied a perennial source of water for rice irrigation. Over the coming decades, climate change is anticipated to affect water supply through its impacts on glacial runoff, or by reducing the level of snowpack (Kerr, 2007). The initial effect would increase water supply; that is, with melting, water availability will increase, starting earlier in the spring, and the likelihood of flooding will increase. As the snowpack diminishes, less water will be available. Increasing severity of droughts, particularly as part of increased desertification globally (IPCC, 2007), is also likely to affect water supply (Figure 9.3).

As water becomes less available, water management for cultural weed control can adjust. For example, rather than transplanting rice into flooded conditions, water savings can be achieved if the field is flooded after the rice crop is sown directly and has reached a certain size. However, in these direct-seeded systems, postemergent weed control is necessary in order to keep weed populations low. In flooded systems, only a handful of weeds, usually those that are tolerant of anaerobic conditions, and resemble the rice plant in the seedling stages (e.g., barnyardgrass) are competitive. This is why flooding is, and has been for

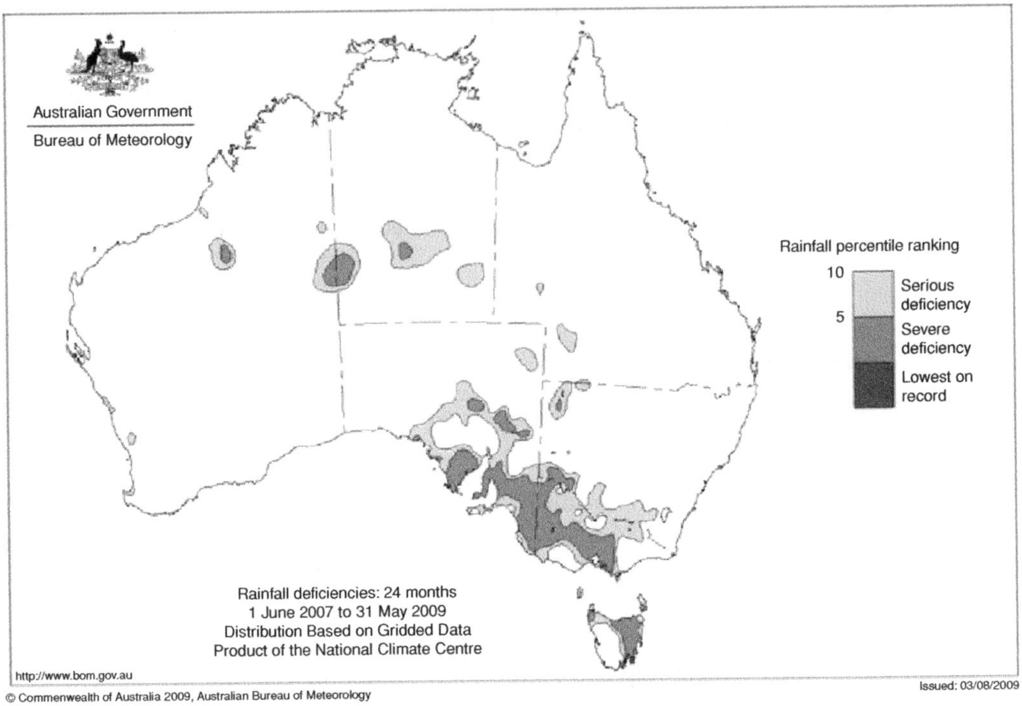

Figure 9.3 Extent of drought in Australia. Climate change is projected to lead to longer and more severe hear waves, droughts and, paradoxically, heavy rainfall events. This will have implications with respect to using flooding as a means of weed control.

millennia, a successful weed control system in rice. Aerobic rice systems, wherein the crop is established via direct-seeding in nonflooded fields, can reduce the costs of labor and water. However, rice is not a good competitor with weeds at the seedling stage and aerobic systems are subject to much higher weed pressure than conventional paddy rice (Singh et al., 2006).

Mulching

Any material spread on the surface that can exclude light can prevent weed seed germination and manage weeds effectively. However, mulches, because of cost, are usually utilized in small areas (e.g., gardens) or for high value crops (e.g., strawberries). Mulching material usually is a natural or man-made substance sufficiently opaque to prevent light from reaching the soil surface, and thick enough to prevent seedling emergence. Examples of effective mulches include straw, sawdust, wood chips, grass clippings, black plastic, and shredded paper. Mulches when used effectively, may also increase soil temperature and conserve water for desired plants.

Living mulches (cover crops or green manure) can also be an effective weed management technique. In these systems, two or more plant species are grown simultaneously. Unlike intercropping where two crops are grown simultaneously, the concept of living mulch is associated with managing row crops in an established plant cover as a means to minimize bare soil and prevent weed seed establishment. The plant cover in turn can be grown as a monoculture or with the cash crop. Such a system has multiple influences including altering temperature, increasing nitrogen (if a leguminous cover crop is used), altering the movement of water, nutrients, and chemicals, improving soil structure, and reducing the establishment and presence of a number of weedy species (Teasdale et al., 2007). The principal disadvantage is the reduced yield from the cash crop as a result of increased competition. As such, a number of cover crops need to be managed carefully to avoid yield reductions. If the cover crop and cash crop exploit different aspects of the soil system (e.g., shallow and deep roots), then little competition usually results (e.g., grass in orchards).

Rising CO_2 and/or climate change could alter the effectiveness of mulches in a number of ways. Generally, mulches are less effective against perennial weeds. That is, food reserves in below-ground structures can permit the growth of perennial plants through even thick mulches. Greater below-ground structures and food reserves with elevated CO_2 may make perennial weeds even more resistant to control by mulches (e.g., Canada thistle). CO_2-induced changes in weed seed production or seed size could also change the effectiveness of mulches. Similarly, increasing climatic extremes could accelerate mulch degradation over time. Conversely, mulches may help to conserve soil during drought, or to help modify temperature extremes, thereby favoring crop over weed species.

Crop Competition and Rotation

Management practices that can help the crop to outcompete the weed can be another useful strategy for weed control. For example, selection of crop lines that are well adapted to a specific set of environmental conditions and that rapidly grow a closed canopy can help prevent weed establishment. If a crop is slower in development, than row spacing can be narrowed to accelerate canopy closure. Other management practices related to planting times, optimum water, and fertilizer use, control of insects, and pathogens, etc., can all contribute to a healthy and vigorous crop stand, with minimal weed competition.

One cultural practice can be especially effective in weed management. The practice of crop rotation, of growing different crops on the same land, from year to year remains a very effective means of weed control. Because many of the most pernicious weeds in any cropping system are those that simulate the life cycle of the crop with respect to emergence, growth pattern, seed dispersal, etc. (e.g., rice and red rice), alternating crops can prevent the accumulation of weed populations that are effective competitors with a given crop species. The more dissimilar alternating crops are with respect to their biology, the less opportunity any individual weed has to become dominant within the weed seed bank. Crop rotation can also allow the use of herbicides or other types of weed management if populations of a given herbicide-tolerant weed become common.

An overview of how rising CO_2 and climate could alter crop–weed competition has already been presented (see Chapter 5). Briefly, there is an emerging consensus that weeds may be better adapted to the kind of rapidly changing environmental conditions represented by climate change (Franks et al., 2007). In addition, there are a number of management decisions that incorporate weed control that may be affected. For example, earlier springs and milder winters could change cropping dates and agricultural management decisions. Fertilizer management, water supply, etc., are also likely to be altered. Land managers will likely be faced with a number of decisions that must be implemented as a means to maintain adequate weed control, with the increasing likelihood that past management strategies for crop establishment may be inadequate to control weeds with increasing carbon dioxide and greater climatic uncertainty.

Allelopathy

Plants (crops and weeds) have been shown in some circumstances to produce secondary chemicals that can have significant effects on the growth of adjacent plants (Olofsdotter and Andersen, 2004). An early study by Putnam and Duke (1974) using 526 accessions of cucumber (*Cucumis sativus*) demonstrated that allelopathic effects against some weed species (broomcorn millet and white mustard) could occur. In their study, one accession inhibited weed growth by 87% and 25 additional lines inhibited growth by 50% or more. Allelopathic effects have been observed for a number of cereal crops including barley (Overland, 1966), rye (Putnam and DeFrank, 1983), rice (Dilday et al., 1998), and wheat (Wu et al., 2000).

Although a number of different plant parts are capable of allelopathic effects, those that are produced in roots or shoots are thought to have the greatest potential value (Putnam, 1985). Although some evidence has suggested that plants may release chemicals directly (e.g., Vivanco et al., 2004) and that this may be part of the mechanism to explain the increased competitive ability of invasive weeds such as spotted knapweed (The "Novel weapons hypothesis"), other researchers have disputed such a role (Duke et al., 2009). At present, the most efficacious means to which allelopathic effects have been employed by land managers is by utilizing residues from known allelopathic plants as mulches prior to planting of the cash crop. For example, wheat or rye can be grown prior to another crop, and following harvest, straw residue may contribute to allelopathic weed control. Management should include a no-till system to allow the residue to remain in sufficiently high concentration, and care should be taken to avoid any allelopathic effects on the crop. In addition, breeding strategies for increasing allelopathic ability of crop plants and efforts to increase our knowledge of the genetic and molecular basis of allelopathy are needed (Olofsdotter and Andersen, 2004).

Changes in temperature, precipitation, and climatic variability are likely to indirectly influence allelopathy by altering rates of decomposition. Because of the cellular and biochemical impacts of carbon dioxide concentration, a reduction in the ratio of carbon to nitrogen with plant material is often observed in response to rising CO_2 levels (e.g., Bazzaz, 1996; Drake et al., 1997). The change in C:N ratio may also affect the concentration of secondary chemicals such as lignin or cellulose. At present, it is unclear if CO_2 or climate induces long-term (multi-year) changes in soil decomposition rates and nitrogen cycling with subsequent effects on the release of allelopathic chemicals. Also, unclear is whether production of (−)-catechin, the phytotoxin found in spotted knapweed could also be affected by CO_2. That is, since catechin is a carbon-based compound, its concentration could increase as atmospheric carbon dioxide increases, stimulating the competitive abilities of a widespread invasive weed. Unfortunately, to date, few studies are available regarding the impact of CO_2 and climate on allelopathy. Shafer et al. (1998) reported no effect of rising CO_2 on competition of tomatoes via allelopathy (phenolic acid production); in contrast, recent results for the exotic biotype of *Phragmites australis* indicated that the gallic acid released by *Phragmites* is degraded by ultraviolet light to produce another toxin, mesoxalic acid, which destroys structural proteins in the roots of susceptible plants (Rudrappa et al., 2009).

Biological Control

Biological control is the planned introduction of natural enemies to lower the population level (or competitive ability) of a weed species (or other pest) to levels that no longer represent an economic or environmental problem. The goal is not to eliminate the pest altogether, since a small number of weeds should be available to maintain the control agent, but to keep populations low enough so that they do not represent a threat to human interests.

Advantages and Disadvantages

Biological control can represent a low-cost, permanent strategy, because once the biocontrol agent is released, it is self-perpetuating and control will continue without human intervention. This differs from chemical weed management, which demands continuous control of the weed population (Figure 9.4). In addition, there are no residues, or long-term chemical release into the environment. Costs may be high initially, but are nonrecurring, and once established no additional costs may be necessary. In addition, biocontrol agents may be able to control a weed over large areas without the use of pesticides.

There are, however, a number of limitations and potential drawbacks to biocontrol. There is a lag period before the biocontrol agent is effective; consequently, this technique is not appropriate for any immediate weed threats. If applied early in the growing season, biocontrol agents are ineffectual, because the weed population(s) may not have built up sufficiently; yet, early competition from weeds during this time may be associated with the greatest yield losses. In addition, biocontrol must be highly selective since any control organism may not be able to distinguish between related plant species. Release of a biocontrol agent without adequate background research and proper precautions can result in the release of a new, potentially devastating pest (e.g., the cane toad, a disastrous case of biocontrol gone awry in Australia). Zimdahl (1993) gives the example of artichoke thistle (*Cynara cardunculus*), a weed in California that is also related to commercial globe artichokes (also *Cynara cardunculus*; Marushia and Holt, 2008). In this instance, any biocontrol agent introduced to control the wild relative would also damage the commercial subspecies. Because specificity of the control agent is so critical, screening of potential agents requires a great deal of time and expense. Hence, biocontrol agents target single weed species and are not effective as broad-scale management tools in crop production, where a number of weeds are usually present.

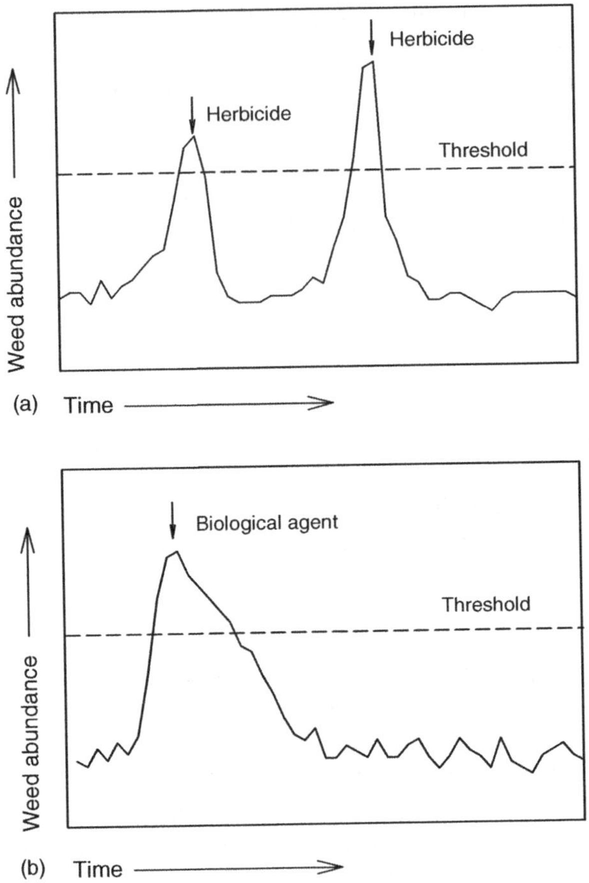

Figure 9.4 Theoretical construct of weed management over time for chemical and biological applications. Although biological control may not be immediate, if successful, it may result in weed populations that remain below an economic or environmental impact threshold over time.

Techniques

There are three techniques of biocontrol based on DeBach's definition of biocontrol as applied by McFadyen (1998) to weeds. These techniques include: (a) "conservation," the protection and inducement of existing populations of organisms thought to have the potential for biocontrol; (b) "augmentation," the practice of introducing biocontrol agents (usually native) after concentrating their numbers; and (c) "classical biocontrol," the importation and release of exotic biocontrol agents. The latter is the basis for most biological weed control currently practiced, whereas conservation is hardly used (McFadyen, 1998).

For augmentation biocontrol, agents such as fungi or insects are manipulated to produce large quantities. The organisms are concentrated and/or packaged and then introduced into the environment when conditions are most favorable for biocontrol. This approach has achieved some commercial success, most notably in the use of fungi usually sold as dried spores with appropriate rewetting agents (e.g., mycoherbicides such as "Collego"). However, while such mycoherbicides may have value in controlled environments, their storage and efficacy requirements make them unsuitable for field conditions (Jackson et al., 1996). Few examples are available that describe the ability or use of native insects for control of indigenous weeds. McFadyen (1998) cites both the timed release of mealy bugs in Australia and South Africa to control Cacti, and the release of the weevil *Cyrtobagous salviniae* to control floating fern salvinia (*Salvinia molesta*) in Australia. Overall, however, the use of augmentation as an economically viable alternative to other forms of weed control remains largely unproven.

At present, classical biocontrol represents an economical and permanent solution when a single weed species poses a major threat and a biological control agent is available. This approach has particular appeal for noxious or invasive weeds that may be spread over a large geographic area. Such weeds may, in fact, be characterized by the lack of natural enemies in their introduced habitat (see Chapter 6). Hence, the transportation of biological antagonists from the weed's original range to the introduced habitat may act as a very effective means of control. Care must be taken, however, to prevent a possible shift in preferential food sources from such an antagonist; that is, any biocontrol agent must be specific to the weed to be controlled, unable to harm desirable plant species, able to survive for long periods of time, and not limited by new enemies in the introduced environment.

Insects have, in a number of cases, provided successful biocontrol of terrestrial weeds. Examples of effective control include the management of prickly pear cactus in Australia in the 1930s through the introduction of the moth borer (*Cactoblastis cactorum*), the utilization of the French chrysomelid leaf beetle (*Chrysolina quadrigemina*) for control of St. John's wort in western rangelands, and the ability of the South American flea beetle (*Agasicles hyrophila*) to keep alligator weed in check in the southern United States. In some instances, control can be achieved with multiple insects as with tansy ragwort, a poisonous weed (Pemberton and Turner, 1990). Research has also been ongoing regarding bacteria and fungi for weed control; however, localized damage rather than widespread control seems to be the result (Ross and Lembi, 1999). Overall, there are numerous efforts currently underway to develop biocontrol of a number of noxious and invasive weed species including kudzu, garlic mustard, purple loosestrife, etc. (www.invasiveplants.net). In many cases, initial biocontrol agents achieve, at best, modest reductions in reproduction of target species, and so multiple agents are introduced over time.

Classical biocontrol may also involve the judicious use of grazing animals (Ross and Lembi, 1999). Geese, goats, cows, etc. have been successful in controlling weeds in some environments. White Chinese geese have been used for grass control in orchards, vineyards, and in some row crops. Geese consume immature grasses and nutsedges and will even prefer large bermudagrass and johnsongrass plants. Cattle have been used to control johnsongrass in fescue pastures, goats have been used to control kudzu; fish (the grass carp, *Ctenopharyngodon idella*) can be used to control hydrilla. The use of grazing animals requires that the animal not be indiscriminate in their feeding, that is, they cannot be turned out and left to themselves; rather, certain husbandry techniques need to be utilized for effective management.

Essential management includes supplemental feeding for essential nutrients, fencing, herding, and sheltering, as well as protection from predators and some degree of health care.

Impact of CO_2 and Climate Change on Biocontrol Efforts

Any plant community represents a multitude of above- and below-ground animal, insect, plant, and microbial interactions. As climate changes, species are likely to respond differentially. However, the use of biocontrol agents (e.g., insects) is dependent on synchrony between various aspects of the plant community. Temperature is a significant abiotic factor in insect ecology and distribution, and a number of studies have shown that earlier springs and warmer temperatures have already altered the geographic range of a number of insect species as well as plant–insect interactions (Montaigne, 2004; Parmesan, 1996; Singer and Parmesan, 1993; Walther et al., 2002). If temperature expands the range of biocontrol agents, this could bring them into contact with other potential plant hosts present at distant locations. This has been suggested by Cannon (1998) who examined the spread of insects in northwestern Europe and the implications for insect invasions in the United Kingdom. While not specific to insects per se, it is worth noting that insect herbivory has increased sharply during previous geological warming periods (Currano et al., 2008). A major concern of climate change therefore involves the timing, presence, and activity of natural or introduced enemies of pernicious weed species. For example, if overwintering of introduced insect populations occurs with warmer winters, but the host weed becomes dormant, this will reduce the survival rates of the insect. Similarly, if the ranges of the host weed and introduced insect vary as a result of fluctuating temperatures, this will have ramifications for the efficacy of biocontrol.

As previously discussed, atmospheric CO_2-induced changes in plant metabolism at the leaf level are likely to include increased C:N ratios, altered concentrations of defensive compounds, changes in carbohydrate and fiber content, and increased water content. At the whole plant level, CO_2-induced changes in allometry and phenology are evident, including time to flowering, flower number, and even pollen production (Springer and Ward, 2007). These changes are likely to alter biocontrol efforts by either altering the feeding behavior and/or fecundity of the bioagent or by altering plant defenses. For example, an increasing C:N ratio in leaves at elevated CO_2 may increase insect feeding rates, but decrease overall fecundity (Coviella et al., 2002; Fajer et al., 1989). CO_2- or climate-induced increases in the toxicology of secondary plant compounds may also complicate weed removal by large animals.

Integrated Weed Control

As sometimes occurs, different weed management techniques can conflict with each other. For example, wood boring insects help control mesquite by increasing their flammability. In contrast, flammability of mesquite is reduced by herbicides that defoliate the tree (Zimdahl, 1993). Integrated pest management (IPM), or integrated weed management (IWM), is a concept designed to prevent such conflicts by coordinating all weed management tools.

It represents a series of monitoring, decisions, controls, and evaluations. First, the action threshold needs to be determined. That is, at what point does the weed population require control? Second, not all weeds (or other biological organisms) require control. Many are innocuous, some may even be beneficial. Consequently, monitoring and identification are critical so that appropriate management decisions are made if an action threshold is reached. This requires proper identification and knowledge of both agronomic and IPS as well as mapping of weed populations and potential damage. This step can assist in determining whether herbicide application is necessary. Third, prevention is emphasized, that is, cultural methods such as rotation, weed-free seed, etc., should be standard practice. Such a strategy is of low-cost and presents little risk to people or the environment. Once monitoring, identification and action thresholds indicate that control is required, then a management decision must be made. Such decisions should reflect knowledge of potential damage, cost, and environmental impact. Effective, less environmentally risky controls, are then chosen initially, for example, mechanical or biocontrol if available. Herbicides may be used in targeted spraying. Control strategies may include a combination of methods to reduce weed populations to acceptable levels. Broadcast spraying of nonspecific chemicals is considered a last resort. Following application, evaluations of the efficacy of the control method, its cost, impacts, etc., should be incorporated into future management decisions.

Changes in climate and carbon dioxide are likely to lead to a myriad of consequences with respect not only to weeds, but to diseases and pathogens as well. The preponderance of evidence to date indicates that there is likely to be an overall change in when and where new insect, pathogens, and weeds will occur. Understanding that some of the current weed and other pest species are likely to decline, and that new weeds may emerge as threats should be recognized as a crucial aspect of preparing for and adapting to climate change. Working to prevent the establishment of new weed species remains a priority for all landowners.

The flexibility and continuum represented by an integrated pest approach may be particularly valuable in helping land managers cope with rising carbon dioxide and climate change. However, there are a number of adaptation strategies to be considered in the context of CO_2 and climate that could increase the overall efficacy of IPM or IWM. For example, land managers may be able to take advantage of adverse seasonal conditions (e.g., drought) that could facilitate weed control. Monitoring for population and incursions of new weed or pest species will be especially critical since weeds and other pests have short generation times, and may adapt quickly to changing environmental conditions. To this end, efforts should be made to increase cooperation across political boundaries (cities, counties, states, and nations) to facilitate early detection of weeds and other pests. If herbicide effectiveness does diminish over time with rising CO_2 and climate, manufacturers and producers may need to review the efficacy and timing of herbicide application within any integrated weed management program. Overall, land managers can prepare for the future by planning for change, utilizing an integrated approach to managing weeds and pests, and anticipating the additional effort needed to manage traditional weeds as well as to increase efforts to identify and eliminate the early establishment of new invasive weed species for their region.

Summary

Strategies to manage the negative consequences of weed impacts include early detection and a number of control measures, including physical, cultural, and biological means of control. While these types of control may not involve the ease of use represented by chemical applications, there are a number of factors, most notably, environmental concerns related to soil and groundwater contamination that make alternative means of weed management attractive even in developed countries. These practices, including chemical use, can be built into an effective, integrated strategy to control not only weeds, but also other pests such as insects and disease. As with chemical control, these practices are likely to be altered by climate change and CO_2, and so management strategies must adjust accordingly. We emphasize the need for greater weed surveillance to prevent establishment, regular reappraisal of the effectiveness of nonchemical and chemical strategies, exploitation of environmental extremes as a potential means of control, and greater overall reliance on integrated management.

In this and previous chapters, we have examined the harm that weeds do and the means by which we can minimize that harm. The following chapter examines the potential benefits offered by weeds in the context of changing climate and increasing carbon dioxide.

References

Bazzaz FA. 1996. *Plants in Changing Environments: Linking Physiological, Population and Community Ecology*. Cambridge University Press, Cambridge, UK, 320 pp.

Bradley BA, Oppenheimer M, Wilcove DS. 2009. Climate change and plant invasion: restoration opportunities ahead? *Global Change Biol.* **15**:1511–1521.

Cannon RJC. 1998. The implications of predicted climate change for insect pests in the UK, with emphasis on non-indigenous species. *Global Change Biol.* **4**:785–796.

Coviella CE, Stipanovic RD, Trumble JT. 2002. Plant allocation to defensive compounds: interactions between elevated CO_2 and nitrogen in transgenic cotton plants. *J. Exp. Bot.* **53**:323–332.

Currano ED, Wilf P, Wing SL, Labandeira CC, Lovelock EC, Royer DL. 2008. Sharply increased insect herbivory during the Paleocene–Eocene thermal maximum. *Proc. Natl. Acad. Sci. U. S. A.* **105**:1960–1964.

Dilday RH, Yan WG, Moldenhauer KA, Gravois KA. 1998. Allelopathic activity in rice for controlling major aquatic weeds. In: Olofsdotter M (ed.) *Allelopathy in Rice*. IRRI, Los Banos, Philippines pp. 7–26.

Drake BG, Gonzalez-Meler M, Long SP. 1997. More efficient plants: a consequence of rising CO_2? *Ann. Rev. Plant Physiol. Plant Mol. Biol.* **48**:609–639.

Duke SO, Dayan FE, Bajsa JN, Meepagala KM, Hufbauer RA, Blair AC. 2009. The case against (−)-catechin involvement in allelopathy in *Centaurea stoebe* (spotted knapweed). *Plant Sig. Behav.* **4**:422–424.

Fajer ED, Bowers MD, Bazzaz FA. 1989. The effects of enriched carbon dioxide atmospheres on plant–insect herbivore interactions. *Science* **243**:1198–1200.

Franks SJ, Sim S, Weis AE. 2007. Rapid evolution of flowering time by an annual plant in response to a climate fluctuation. *Proc. Natl. Acad. Sci. U. S. A.* **104**:1278–1282.

Gerhards R, Christensen S. 2003. Real-time weed detection, decision making and patch spraying in maize, sugar beet, winter wheat and winter barley. *Weed Res.* **43**:385–392.

Graham J, Simpson A, Crall A, Jarnevich C, Newman G, Stohlgren T. 2008. Vision of a cyber-infrastructure for nonnative, invasive species management. *Biosciences* **58**:263–268.

Hill GD. 1982. Impact of weed science and agricultural chemicals on farm productivity in the 1980s. *Weed Sci.* **30**:426–429.

IPCC: Climate Change. 2007. *Impacts, Adaptation and Vulnerability*. IPCC Secretariat, Geneva, Switzerland, 986 pp.

Jackson MA, Schisler DA, Slininger PJ, Boyette CD, Silman RW, Bothast RJ. 1996. Fermentation strategies for improving the fitness of a bioherbicide. *Weed Technol.* **10**:645–650.

Kerr RA. 2007. Global warming coming home to roost in the American West. *Science* **318**:1859–1860.

Lawrence RL, Wood SD, Shelley RL. 2005. Mapping invasive plants using hyperspectral imagery and Breiman Cutler classifications (randomFOrest). *Remote Sens. Environ.* **100**:356–362.

Marushia RG, Holt JS. 2008. Reproductive strategy of an invasive thistle: effects of adults on seedling survival. *Biol. Invas.* **10**:913–924.

McFadyen R. 1998. Biological control of weeds. *Annu. Rev. Entomol.* **43**:369–393.

Melander B, Rasmussen IA, Barberi P. 2005. Integrating physical and cultural methods of weed control: examples from European research. *Weed Sci.* **53**:369–381.

Montaigne F. 2004. The heat is on: eco-signs. *Nat Geo.* **206**:34–55.

Morgan JA, Milchunas DG, LeCain DR, West M, Mosier AR. 2007. Carbon dioxide enrichment alters plant community structure and accelerates shrub growth in the shortgrass steppe. *Proc. Natl. Acad. Sci. U. S. A.* **104**:14724–14729.

Olofsdotter M, Andersen S. 2004. Improvement of allelopathy in crops for weed management-possibilities, breeding, strategies and tools. In: Interjit (ed.) *Weed Biology and Management*. Kluwer Academic Publishers, Dordrecht, The Netherlands, pp. 317–328.

Overland L. 1966. The role of allelopathic substances in the smother crop barley. *Am. J. Bot.* **53**:423–432.

Parker-Williams AE, Hunt ER Jr. 2004. Accuracy assessment for detection of leafy spurge with hyperspectral imagery. *J. Range Manag.* **57**:106–112.

Parmesan C. 1996. Climate and species range. *Nature* **382**:765–766.

Pemberton RW, Turner CE. 1990. Biological control of *Senecio jacobaea* in northern California, an enduring success. *Entomophaga* **35**:71–77.

Polley HW, Mayeux HS, Johnson HV, Tischler CR. 1997. Viewpoint: atmospheric CO2, soil water, and shrub/grass ratios on rangelands. *J. Range Manage.* **50**:278–284.

Putnam AR. 1985. Weed allelopathy. In: Duke SO (ed.) *Weed Physiology*, Vol. I. CRC Press, Boca Raton, FL, pp. 131–155.

Putnam AR, DeFrank J. 1983. Use of phytotoxic plant residues for selective weed control. *Crop Prot.* **2**:173–181.

Putnam AR, Duke WB. 1974. Biological suppression of weeds: evidence for allelopathy in accessions of cucumber. *Science.* **185**:370–371.

Rejmanek M. 2000. Invasive plants: approaches and predictions. *Aust. Ecol.* **25**:497–506.

Rogers HH, Runion GB, Krupa SV. 1994. Plant responses to CO_2 enrichment with emphasis on roots and the rhizosphere. *Environ. Pollut.* **83**:155–189.

Rosenzweig C, Hillel D. 1998. *Climate Change and the Global Harvest*. Oxford University Press, New York, 324 pp.

Ross MA, Lembi CA. 1985. Applied Weed Science. Burges Publishing CO., Minneapolis, MN, 474 pp.

Ross MA, Lembi CA. 1999. Methods of weed control. In: Ross MA, Lembi CA (eds) *Applied Weed Science*. Prentice Hall, Saddle River, NJ, pp. 48–75.

Rudrappa T, Choi YS, Levia DF, Legates DR, Lee KH, Bais HP. 2009. Phragmites australis root secreted phytotoxin undergoes photo-degradation to execute severe phytotoxicity. *Plant. Signal. Behav.* **4**:506–513.

Sasek TW, Strain BR. 1990. Implications of atmospheric CO_2 enrichment and climatic change for the geographical distribution of two introduced vines in the USA. *Climate Change* **16**:31–51.

Shafer SR, Blum U, Horton SJ, Hesterberg DL. 1998. Biomass of tomato seedlings exposed to an allelopathic phenolic acid and enriched atmospheric carbon dioxide. *Water Air Soil Poll.* **106**:123–136.

Singer MC, Parmesan C. 1993. Sources of variation in patterns of plant–insect association. *Nature* **361**:251–253.

Singh L, Bhushan JK, Ladha RK, Gupta AN, Sivaprasad B. 2006. Weed management in dry-seeded rice (*Oryza sativa*) cultivated in the furrow-irrigated raised-bed planting system. *Crop Prot.* **25**:487–495.

Springer CJ, Ward JK. 2007. Flowering time and elevated atmospheric CO_2. *New Phytol.* **176**:243–255.

Sutherst RW, Maywald GF, Yonow T, Stevens PM. 1999. *CLIMEX: Predicting the Effects of Climate on Plants and Animals*, Vol. 1.1. CSIRO, Melbourne, Australia.

Teasdale JR, Coffman CB, Magnum RW. 2007. Potential long-term benefits of no-tillage and organic cropping systems for grain production and soil improvement. *Agron. J.* **99**:1297–1305.

Vincent C, Panneton B, Fleurat-Lessard F. 2001. *Physical Control Methods in Plant Protection*. Springer Press, Heidelberg, Germany, 329 pp.

Vivanco JM, Bais HP, Stermitz FR, Thelen GC, Callaway RM. 2004. Biogeographical variation in community response to root allelochemistry: novel weapons and exotic invasion. *Ecol. Lett.* **7**:285–292.

Walther GR, Post E, Convey P, Menzel A, Parmesan C, Beebee TJ, Fromentin JM, Hoegh-Goldberg O, Bairlein F. 2002. Ecological responses to recent climate change. *Nature* **416**:389–395.

Watson AK. 2002. Flooding: weed control in rice. In: Pimental DT (ed.) *Encyclopedia of Pest Management*. Taylor and Francis, On-line, DOI: 10.1081/E-EPM-100001003.

Webb M, Conroy C. 1995. The socio-economics of weed control on small-holder farms in Uganda. In: *Proceedings, Brighton Crop Protection Conference—Weeds*. British Crop Protection Council, Farnham, UK, pp. 157–162.

Wu H, Pratley J, Lemerle D, Haig T. 2000. Evaluation of seedling allelopathy in 453 wheat (*Triticum aestivum*) accessions against annual ryegrass (*Lolium rigidum*) by the equal-compartment-agar method. *Aust J. Agric. Res.* **51**:937–944.

Zimdahl, RL. 1993. *Fundamentals of Weed Science*. Academic Press, New York, 454 pp.

Ziska LH. 2003. Evaluation of the growth response of six invasive species to past, present and future carbon dioxide concentrations. *J. Exp. Bot.* **54**:395–404.

10 Benefits from Weeds

"Weeds are flowers too, once you get to know them."

A.A. Milne

Introduction

One of the most pernicious weeds of riparian landscapes, particularly in the southwestern United States, is giant reed, or *Arundo donax*. Much time and effort has been expended in its control and removal, but even a small vegetative piece floating downstream can establish itself into a weedy copse growing up to 10 m in height. Yet, *Arundo donax* exemplifies the paradox of some weed species in that while particularly undesirable from a landscape point of view, it is also the plant source for musical reeds for all woodwind instruments (Figure 10.1).

This illustrates the occasional paradox of what constitutes a weed. As we have shown thus far, weeds are plants that can do tremendous harm to both the natural and managed environment. But from an ethnobotanical perspective, some weeds can, in certain circumstances and cultures, be useful.

In this chapter, we will explore the current and potential uses of weeds, both in regard to food and feed, as medicines, their use in agroecosystems, as a means to help control diseases or insects, in preventing eutrophication and as potential biofuels. We hope to emphasize not only how climate change may alter their benefits, but also how those specific benefits could be adapted to or potentially mitigate the impacts of climate change.

Feed and Food

For animals, some weeds can be a valuable source of protein. For example, redroot pigweed, common lambsquarters, and common ragweed have fiber and crude protein equal to that of alfalfa hay when alfalfa and weeds were harvested at the same growth stage (Marten and Andersen, 1975; Temme et al., 1979). Marten et al. (1987) reported similar results regarding the forage quality of nine perennial broadleaf and grass weeds relative to alfalfa and bromegrass. A similar study on invasive weeds in Idaho (rush skeletonweed, meadow hawkweed, houndstongue, sulfur cinquefoil, yellow starthistle, Dalmatian toadflax, hoary

Weed Biology and Climate Change, Lewis H. Ziska and Jeffrey S. Dukes, © 2011 Blackwell Publishing Ltd.

(a) (b) (c)

Figure 10.1 Picture of Arundo donax: (a) an invasive weed in the western United States. However, Arundo also supplies material to make single and double reeds (b) for all woodwind instruments. At right (c) Dr. Marguerite Levin of Towson University plays the bass clarinet. (Arundo photo courtesy of Michael Rauterkus of the University of California, Riverside; Clarinet photo is courtesy of Dr. Marguerite Levin of the Music Department of Towson University.)

cress, and tansy ragwort) found that these invasive weeds also have forage values similar to many native plants and could provide reasonable nutrition for livestock (Frost et al., 2008), although some of these species produce toxins and/or physical structures such as spines that severely limit their appeal. Kudzu, the archetypal invasive vine, can produce forage that is high in quality and palatability (Terrill et al., 2009). Kudzu hay may have a crude protein content of 15–18% and a total digestible nutrient value of over 60% (dry matter basis) which compares favorably to other leguminous forage hays (e.g., alfalfa). However, there are some caveats. For example, kudzu quality decreases as the amount of large vines increases relative to leaves. In addition, kudzu may be difficult to cut and bale as hay because of its viny growth habit. Overall, it was unclear if kudzu or other invasive species would, in fact, contain any additional secondary compounds related to actual palatability and/or toxicity, or if they would be preferentially selected by grazing animals.

In general, as previously discussed (see Chapter 3), rising CO_2 can reduce the nitrogen content of foliage, in part because the need for nitrogen in carbon fixation is reduced. Consequently, as CO_2 rises, the protein content of weed forages could also decline with subsequent effects on palatability. However, leguminous weeds, such as kudzu, may be less impacted by the change in CO_2. When grown under elevated atmospheric carbon dioxide (CO_2), leaf nitrogen content decreases less for legumes than for nonlegume C_3 plants (Karowe, 2007). Hence, kudzu may be more attractive as a forage source. In addition, with greater variation in precipitation, and the threat of prolonged drought, the use of kudzu may be advantageous.

For example, a few Alabama farmers still harvest kudzu as hay by annual or biennial harvesting. It is most frequently used in this manner during dry periods, since its deep root system allows it to obtain moisture for growth when other forage species may be adversely affected.

There are only about 200 or so plant species (out of a total exceeding 250,000) that are directly consumed by humans, and among these, a handful of cereals such as rice, wheat, and corn, that supply the bulk of the caloric intake. There is interest then in looking at other potential food sources, particularly in a population that will soon exceed 7 billion.

Some plant species that are now considered weeds in the modern United States have been used as sources of food in other cultures and times. For example, young pigweed and lambsquarters plants were used by pioneers crossing the west. Neolithic, Bronze Age, and early Iron Age people ate it, and the Romans and Europeans used it extensively (Zimdahl, 1993). Lambsquarters is related to spinach and many cultures regard lambsquarters as one of the most delicious of wild vegetables. Leaves of other weedy annuals such as shepherd's purse and purselane can also be eaten as greens. (Vineland, New Jersey, regards itself as "the dandelion capital of the world" and celebrates the uses of dandelion in everything from jewelry to casseroles.) Duckweed, among the worst aquatic weeds, has the potential to help world hunger in part because it is rich in protein and highly abundant (Leng, 1999). It is used currently in Thailand as a food source in soup, and has a very high potential to feed livestock. For example, 10 acres (4 hectares) could supply the nutritional needs of 100 diary cows for 1 year (Culley et al., 1981). The use of weeds as potential sources for new crop domestication may offer significant ecological advantages in an uncertain climate. For example, some might be able to grow in marginal soils that might otherwise be unsuitable for agriculture; they could provide opportunities to use saline soil for agriculture, or for polluted areas, or extreme temperatures or droughts. However, as with all introduced species, extreme care must be taken to ensure that any new plant species does not transform into a dangerous weed pest.

There is also the issue of increasing diversity within current food sources. Even though all individuals of a species may have the same set of genes, they have different versions, or alleles, of these genes. Crop varieties or cultivars differ from each other because they have different alleles and because they have different control sequences that determine which genes are used and to what extent. This is why there are so many different varieties of apples, tomatoes, and rice, etc.

Preservation of crop diversity is necessary because each adaptation represents a different response to a range of abiotic (e.g., temperature and water) and biotic (insects and diseases) stresses. A single variety over long periods of time will encounter viruses or bacteria or environmental limitations that will result in yield vulnerability. Therefore, having additional varieties of the same crop gives farmers a "tool" to begin to adapt to changing conditions. For example, wild relatives of potato were used to derive new lines to combat the plague of *Phytophera infestans* that infected the potato fields of Ireland and killed over a million people in 1845. When *grape phylloxera* (*Daktulosphaira vitifoliae*) infestations threatened European grapevines (*Vitis vinifera*), wild grape rootstocks (*Vitis rotundifolia*) that were resistant to this insect were sent to France from Oklahoma and Texas, saving the French wine industry.

Given that many of the worst weeds are wild relatives of crops (e.g., wild oats and oats, sorghum and shattercane), can they be used as a unique source of genes to begin to adapt cultivated lines to climate change? In Chapter 4, we noted that the differential response of cultivated and weedy (red) rice to elevated CO_2 (Ziska and McClung, 2008) could impact crop/weed competition. But could the wild rice biotype also be superior with respect to characteristics that could be advantageous with climatic uncertainty? We compared the wild, red-rice line "Stuttgart-S" to the cultivated rice line, "Clearfield 161" to determine the potential change in seed yield per plant over the CO_2 increase that occurred (approximately) during the twentieth century.

Surprisingly, Stuttgart-S showed the greatest relative response. However, Stuttgart has one major drawback, the seed shatter. Any use of wild rice (or any wild grain) therefore would necessitate adding the nonshattering mutation. In addition, research would be needed to pattern wild grains into cultivated crops that would have other desired characteristics such as resistance to pathogens (Figure 10.2).

Because wild lines have so much genetic variability, adaptation of current crops to climatic extremes using the wild lines' genetic material is an attractive possibility. For example, as temperatures increase, reproduction often fails before vegetative growth (Hatfield et al., 2008). This is true for cultivated rice as well. In contrast, some wild rice lines flower earlier in the morning or at night and therefore escape potential reproductive stress (Prasad et al., 2006). Overall, the initial results of Ziska and McClung (2008) indicate a greater physiological plasticity and genetic diversity among wild rice biotypes relative to cultivated rice varieties. Potentially, this greater variation in rice and other cereals may also provide a

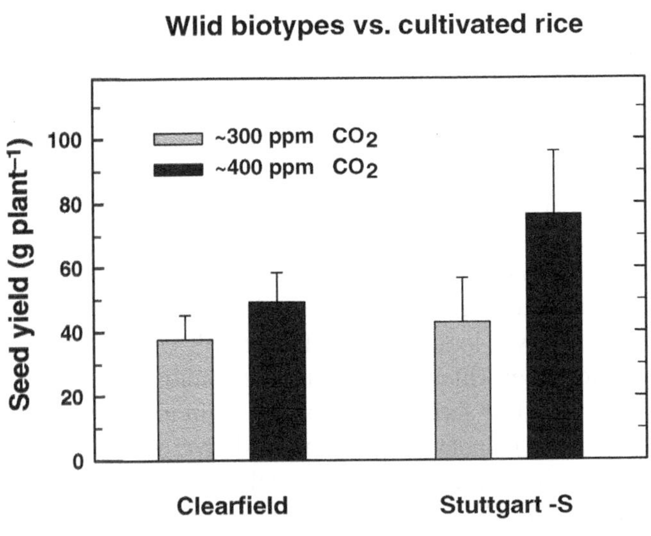

Figure 10.2 Seed yield per plant of cultivated ("Clearfield") and weedy red (Stuttgart-S) rice in response to the change in atmospheric carbon dioxide during the twentieth century.

unique genetic resource that could be incorporated into new crop cultivars in order to increase their adaptability to CO_2 and/or climatic change. Consequently, it is worth preserving the genetic variability of wild crop relatives (i.e., weeds), particularly in habitats where climate change may put crops at risk.

Medicinal Uses

Plants manufacture a wide range of secondary chemicals that defend their tissues against viral diseases, fungal pathogens, and herbivores. Fortunately for us, many of these secondary compounds also affect human physiology. The use of selected plant species and their associated chemical compounds as remedies for ailments is ubiquitous throughout human culture (e.g., Schultes and Reis, 1995). Diversity in the production of secondary chemical products remains an important source of existing and new metabolites of pharmacological interest (Table 10.1). Even in developed countries, where synthetic drugs have replaced herbal medicines, 25% of all prescriptions dispensed from community pharmacies from 1959 through 1980 contained plant extracts or active principles prepared from higher plants (e.g., codeine; Farnsworth et al., 1985). For developing countries, however, the World Health Organization (WHO) reported that more than 3.5 billion people, or approximately half of the world's population, rely on plants as components of their primary health care (WHO, 2002).

Table 10.1 A partial list of plant-derived pharmaceutical drugs and their clinical usage. Although many of these drugs are synthesized in developing countries, the World Health Organization estimates that as many as 3.5 billion people still rely on botanical sources for medicines (WHO, 2002).

Drug	Action/clinical use	Plant species
Acetyldigoxin	Cardiotonic	*Digitalis lanata* (foxglove)
Allyl isothiocyanate	Rubefaciant	*Brassica nigra* (black mustard)
Artemisinin	Antimalarial	*Artemisia annua* (sweet Annie)
Atropine	Antichotinergic	*Datura stramonium* (jimsonweed)
Berberine	Bacillary dysentery	*Berberis vulgaris* (barberry)
Codeine	Analgesic	*Papaver somniferum* (poppy)
D-Pinitol	Expectorant	Various species
L-Dopa	Anti-Parkinson	*Mucuna pruriens* (velvetbean)
Ephedrine	Antihistamine	*Ephedra sinica* (Mormon tea)
Galanthamine	Cholinesterase inhibitor	*Lycoris squamigera* (surprise lily)
Kawain	Tranquilizer	*Piper methysticum* (kava)
Lapachol	Anticancer, antitumor	*Tabebuia avellandedae* (lapacho tree)
Ouabain	Cardiotonic	*Strophanthus gratus* (climbing oleander)
Quinine	Antimalarial	*Cinchona ledgeriana* (Peruvian bark)
Salicin	Analgesic	*Salix alba* (willow)
Taxol	Antitumor	*Podophyllum peltatum* (mayapple)
Vasicine	Cerebral stimulant	*Vinca minor* (periwinkle)
Vincristine	Antileukemic agent	*Catharanthus roseus* (Madagascar periwinkle)

Wild plants, including a number of familiar weed species, are recognized for their curative powers. Plants with the word "officinale" in their species name were included in a list of drugs during medieval times (i.e., the Latin translation of offinalis is "of use in the apothecary"); "wort," another common Latin suffix in plant names means healing, whereas a plant with "bane" added to its common name was recognized as "protection." Night-shade (*Atropa bella-donna*) is a common weed of potato and produces a wide range of tropane alkaloids. Atropine derived from this plant was at one time used by chic young women to enhance pupil dilation (a physiological sign of sexual arousal), hence the species name, *bella-donna* ("beautiful lady" in Latin). Jimson weed produces scopolamine, an analog of "ecstasy" (3,4-methylenedioxymethamphetamine), used by the ancient Greeks to relieve the pain of childbirth. Among medicinal plants, the therapeutic uses of opiate alkaloids from poppy (*Papaver* spp.) have long been recognized (e.g., Sumerians at the end of the third millennium BC recognized the analgesic properties of poppy and named it *hul gil* or "joy plant").

Perhaps one of the most intriguing weeds in medicine is sweet Annie (*Artemisia* species), an annual herb native to China, where it is known as *qinghao*, but found throughout the eastern United States as a roadside weed. *Artemisia annua* contains artemisinin, a compound that has a known antimalarial effect by suppressing Plasmodium's ability to use host erythrocyte protein (Posner et al., 1999). Artemisinin and related compounds have become essential in malaria case management (WHO, 2001).

Because of their importance in human health, there are a number of studies that have begun to examine the potential impact of rising CO_2 and related aspects of climatic change on the production of these secondary compounds. An increase in the antidepressants' hypericin and pseudohypericin was noted in St. Johnswort at a CO_2 concentration of 1,000 ppm relative to ambient CO_2 conditions (Stuhlfauth and Fock, 1990; Zobayed and Saxena, 2004).

Jimson weed and wild tobacco were also studied to determine CO_2- and temperature-induced changes in three secondary alkaloids, nicotine, atropine, and scopolamine (Ziska et al., 2005). Atropine and scopolamine are known as strong anticholinergics. That is, they act as antagonists to the transmission of nerve impulses by acetylcholine. Therapeutic uses of atropine, in addition to pupil dilation, include stimulation of heart rate in patients suffering from bradycardia and as an antidote to G- and V-type nerve agents (soldiers on active duty are given an atropine "stick" or hypodermic in case of exposure). Scopolamine is used as a sedative prior to anesthesia and as an antispasmodic in certain disorders characterized by restlessness and agitation (e.g., Parkinsonism). It can also be prescribed in transdermal patches to prevent motion sickness. The addictive negative effects of nicotine are well known, less known are its role in replacement therapy for tobacco cessation, ulcerative colitis, and potential therapeutic uses in cognitive disorders such as Parkinson's and Alzheimer's (Rezvani and Levin, 2001; Yang et al., 2003). The Ziska et al. (2005) study demonstrated that CO_2 and temperature separately and concurrently had significant effects on the concentration and production of all alkaloids from both plant sources. Recently, the same group also quantified changes in the growth and alkaloid content of wild poppy (*Papaver setigerum*, a noted weed in commercial *P. somniferum* production) to both recent and projected increases in atmospheric carbon dioxide. The experimental values (300, 400, 500, and 600 ppm CO_2) corresponded roughly to the atmospheric concentrations that existed in the middle of the

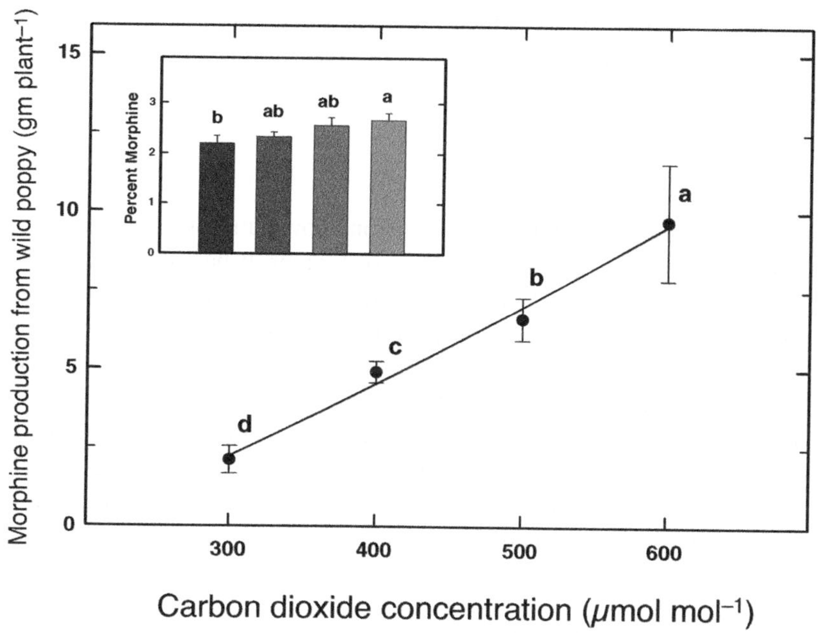

Figure 10.3 Changes in morphine production and concentration (mean ± SE) from wild poppy (*Papaver setigerum*) as a function of recent and projected increases in atmospheric carbon dioxide. Different letters indicate significant differences as a function of CO_2 concentration using Fisher's protected least significant difference.

twentieth century, the current concentration and near and long-term projections for the current century (Ziska et al., 2008). Increasing CO_2 from 300 to 600 ppm increased the number of capsules, capsule weight, and latex production by 3.6, 3.0, and 3.7 times, respectively, on a per plant basis relative to the 300 ppm baseline. Quantification of secondary compounds included morphine, codeine, papaverine, and noscapine. The amount of all alkaloids increased significantly on a per plant basis, with the greatest relative increase occurring with recent increases in atmospheric carbon dioxide (e.g., from 300 to 400 μmol mol^{-1}). Morphine production also increased in response to rising CO_2 levels (Figure 10.3). Overall, these data suggest that both CO_2 and temperature are likely to alter the concentration and production of pharmacologic compounds derived from plants that in certain circumstances are considered weeds. As climate changes, where and in what abundance such plants will grow, is also likely to be in flux.

Agroecological Benefits from Weeds

While the impacts of weeds on natural and managed plant communities are well known, and their control emphasized, there is also a school of thought that advocates using some weed species as beneficial or protective biota of managed ecosystems. Jordan and Vatovec

(2004) have integrated many of these ideas, and see weeds as playing three principal roles: (1) improving soil quality and fertility; (2) as "buffers" to negate or modify insect and/or disease damage; (3) cover crops.

Soils

In certain conditions, weeds can reduce soil erosion, slow nutrient loss, increase soil organic matter, improve nitrogen content, and conserve soil water. It has been suggested that some weeds with deep root systems may be able to transfer nutrients from deep soil layers to depths that can be accessed by desirable crop species (Jordan and Vatovec, 2004). Furthermore, such deeper rooted weeds may be valuable in reducing erosion. Patriquin (1986) has also suggested that weeds may make a significant contribution to nitrogen cycling; in a study of the nitrogen budget on a farm site, it was noted that weeds during fallow periods kept 10 to 20 kg N ha^{-1} from leaching off the soil. This may be particularly important in agricultural land where economic costs may limit N input. Swamy and Ramakrishnan (1988) also note that weeds are early colonizers after slash and burn agriculture and may help accumulate nutrients prior to planting the crop.

Weeds can also play an important ecosystem role in their associations with beneficial soil biota. As bare soil is colonized by plants (usually weeds), there is a concomitant increase in soil microbes (Wardle, 1992). Certain weeds are known to host arbuscular mycorrhizal fungi (AMF; Jordan et al., 2000). AMF have a number of benefits from an agroecosystem (crop) perspective, including improved water relations, pest and disease resistance, tolerance to root pathogens, and enhanced nutrient uptake, particularly phosphorous (Dodd, 2000). The density and diversity of AMF appears to be, in turn, directly related to weed diversity (Rabatin and Stinner, 1989); that is, systems with higher levels of weed biomass are associated with a rich and diverse soil biota (Altieri, 1991). There has also been the suggestion that certain weed–AMF symbionts may be able to function as nutrient redistributors within a cropping system, transferring nutrients from dying or senescing plants to others that are at an earlier stage of growth. If so, planned senescence of weeds, either by sublethal doses of herbicide, or timing spring plantings to coincide with the demise of winter annual weeds, could be beneficial. At present, a preliminary study suggests that increasing CO_2 could benefit grass and legume monocultures (Gamper et al., 2004).

Buffers

Weeds may also act as buffers or modifiers of different processes within agroecosystems, particularly herbivory. Modern agriculture consists of high-yielding monocultures, and the reduction in plant diversity has detrimental effects on beneficial insects. In cropping systems, lack of alternative sources of food, shelter, hibernation sites, or breeding areas can have negative effects on native predators of crop pests (Altieri and Whitcomb, 1979). Plant species other than the crop can provide alternative habitats in crop monocultures that allow for a more diverse and abundant complex of invertebrate predators; predators that can be effective in regulating populations of insect herbivores (Barbosa and Wratten, 1998). One strategy to regain greater diversity in crop fields may be to manipulate the composition of the weed flora. Weed borders, occasional weedy strips, or weeds at certain times in the crop growth cycle may help to increase diversity. By doing so, weeds can contribute to ecological control by affecting the spatial and temporal distribution of plant diversity and abundance. Some weeds may function as a refuge for natural enemies of crop pests, others as an alternate source of food for the pests themselves. Weeds may add diversity to the landscape that visually or olfactorally confuses pests, or they may create a less hospitable microclimate for pests (Bugg, 1992). Jordan and Vatovec (2004) illustrate how common knotweed can be managed spatially and temporally in permanent vegetable production as a means to increase diversity and reduce crop pests (Bugg et al., 1987). When grown with cultivated grapes, wild blackberries act as a host for a leafhopper that hosts a parasite of the grape leafhopper (*Erythroneura elegantula*). Some Japanese farmers graft tomato scion onto rootstock of weedy nightshades (*Solanum* spp.) to avoid root disease.

Cover "Crops"

Certain weeds may have potential benefit as "self-sowing" cover crops in certain management situations. Cover crops have a number of agroecosystem benefits related to soil conservation, water retention, increased organic matter, and pest management. Jordan and Vatovec (2004) describe certain early-emerging weeds that could act to suppress later emerging problematic weed species. In this circumstance, these weeds, along with a planned cover crop, were thought to significantly control the number of later emerging weed species that might otherwise be difficult to control. They argue that the ability of weeds to self-germinate in a variety of different soils and environmental conditions would be advantageous in alleviating some of the difficulties in the timely establishment of intentional cover crops. They also emphasize that certain weeds might be valuable components as part of a multispecies cover-crop polyculture that can provide long-term sustainable benefits for the cropping system.

Modifications with Climate Change

If precipitation patterns change, with greater occurrence of extreme events, preventing erosion may become more challenging. Some weed species may find beneficial use as soil stabilizers. On the other hand, some of the worst weeds of the United States (such as kudzu, multiflora rose, and Lehmann's lovegrass) became established after being introduced to stabilize soil. Many of these plants were promoted by the U.S. Soil Conservation Service (now the Natural Resources Conservation Service).

Phytoremediation

There are a number of very pragmatic uses for weeds in modifying or remediating anthropogenic pollution. For example, because water hyacinth can accumulate up to 2,000 times the concentration of heavy metals such as selenium, manganese, and chromium in surrounding water, it can be used as a bioindicator for these metals. Water hyacinth can also be used to remove excess nutrients from water and can reduce eutrophication (Rogers and Davis, 1972).

Eutrophication refers to the accumulation of excess nutrients in bodies of water that stimulates excessive plant growth, reducing the amount of dissolved oxygen in the water when dead plant material decomposes. Under optimal conditions, a hectare of water hyacinth could absorb the daily nitrogen and phosphorus waste of over 800 people. If water hyacinth is then collected and dried, it can be used for a variety of purposes, including as an additive to corn silage, as an energy feedstock, or as a means to supplement soil organic matter. Similarly, the bulrush (*Scirpus* spp.) has been used by tribesman along the Nile to remove pollutants from water (Zandstra and Motooka, 1978). One of the most intriguing uses of weeds for phytoremediation was the use of mustards and lambsquarters in New Orleans soil following Hurricane Katrina (Thomas, 2006). In this instance, fast-growing plants (e.g., weeds) were used to absorb a wide range of toxins, including lead, arsenic, and pesticides from the city soil. This strategy is considerably cheaper than removing the soil and having to filter or physically remove all of the contamination. The use of plants to neutralize toxins in soil has been sanctioned by the Environmental Protection Agency (EPA), and phytoremediation has, to date, been utilized on approximately 50 superfund clean-up sites.

The impact of rising CO_2 and/or climate change on such phytoremediation efforts is unclear. Uptake of certain materials may be proportional to plant transpiration, and at the leaf level would be negatively impacted by rising CO_2 and drought; alternatively, CO_2-induced larger plants could accumulate greater amounts of certain toxins. For plants such as water hyacinth that are distributed in part, by temperature, warmer weather could extend its northward range. However, since this species is also seen as an aquatic weed, land managers might view this more as a curse than a benefit.

Biofuels

At this time, there is tremendous interest in decreasing dependency of the United States economy on foreign oil. Using "green" sources of fuel has tremendous appeal, both with respect to limiting the production of greenhouse gases and as a means to energy independence. In pursuit of these goals, a number of studies have focused on using plant-based ethanol, either as carbohydrate (starch or sugar), or as cellulose. Cellulosic ethanol has received a great deal of attention and involves the removal of cellulose from its lignin "prison" through the use of pretreatment techniques including acid hydrolysis, steam explosion, ammonia fiber expansion, alkaline wet oxidation, and ozone pretreatment. While industrial cellulosic ethanol may be widespread in the future, at present, most ethanol is derived from starch-based sources, primarily corn. Since corn is one of the three principal cereals (i.e., corn, wheat, and rice) that supply 50% of the world's calories, there is increasing concern that greater diversion of corn feedstocks to meet ethanol demand may contribute to rising food prices and global hunger (Msangi et al., 2007). There is also considerable controversy regarding the net energy balance of ethanol from corn. For example, research from the USDA's Economic Research Service (Shapouri et al., 1995) indicated an energy ratio (i.e., energy output divided by energy input) of 1.24 for corn ethanol, whereas reports from Cornell (Pimental and Patzek, 2005) indicated that corn required 29% more fossil energy than is contained in the ethanol produced. In addition, rather than offsetting greenhouse gas emissions, recent evaluations of land expansion needed to meet the future demand for corn-based ethanol suggest extensive conversion of existing forest and grasslands, with a subsequent net increase in the production of greenhouse gases (Searchinger et al., 2008).

As large-scale capacity for producing cellulosic ethanol is developed, researchers continue to evaluate potential feedstocks. Candidate species include a number of potential weed species including *Miscanthus*, Giant reed, Russian thistle, castor bean, and mesquite. Obviously, any use of these weeds would have to be carefully regulated, however, some weed species already persist in such large geographic areas that use of existing populations as biofuels could both eliminate the weed and generate energy.

One such weed may be kudzu, among the quintessential examples of an invasive weed. Kudzu currently infests 3 million hectares in the United States and is increasing its areal coverage by over 50,000 hectares per year (Forseth and Innis, 2004). Kudzu costs the U.S. economy some $500 million per year in the form of lost crop and forest productivity, control costs, and property damage (Blaustein, 2001). As a result of its aggressive growth, kudzu overtops all forms of vegetation, and frequently forms a dense monoculture that destroys habitat for most native plants and animals. Along its southern tier of infestation from Louisiana to Georgia, "Kudzu-scapes" are large blankets of kudzu that cover all existing vegetation. Kudzu can also alter regional air quality by emitting isoprene (Sharkey and Loreto, 1993), and can reduce water quality by fixing large amounts of nitrogen that leaches into surrounding watersheds (Forseth and Innis, 2004). Because of these negative aspects, kudzu has earned a notorious reputation in the United States, where it has been referred to as "the plant that ate the South," and "a vegetal form of cancer" (Shurtleff and Aoyagi, 1985).

The success of kudzu is due in part to its production of large storage roots. These can exceed 10 cm in diameter and extend over 2 m in length, weighing as much as 180 kg (Blaustein, 2001; Miller and Edwards, 1983). Kudzu roots are rich in starch (up to 25% of root content; Tanner and Hussain, 1979). In Asia, the roots are harvested for this starch, which is used as a thickening agent in food preparation and the plant is cultivated on a small scale (Shurtleff and Aoyagi, 1985). The vines and roots also produce high-quality fiber for paper and textiles (Parks et al., 2002). In North America, people have suggested using kudzu for paper and textile production, starch production, and even fuel sources; however, few if any attempts have been made to exploit such suggestions (Parks et al., 2002; Tanner and Hussain, 1979; Tanner et al., 1979). In nature, the starch in kudzu's roots helps the vine grow quickly in the spring, allowing for the rapid development of a dense canopy that shades competing vegetation (Forseth and Innis, 2004).

In response to the high price of petroleum and new initiatives by governments to reduce dependence on imported fossil fuels, the demand for bioethanol feedstocks in the Americas and Europe has risen dramatically in recent years. In 2006, 17% of the U.S. corn output

Table 10.2 Carbohydrate content of kudzu organs harvested in December 2006 in Alabama and October 2006 in Maryland, the United States.

	Lee County, AL		Rockville, MD	
Plant part	μmol g^{-1}	% of dry matter	μmol g^{-1}	% of dry matter
		Glucose		
Leaf	11 ± 12	0.2	84 ± 13	1.5
Vine	9 ± 1	0.2	8 ± 2	0.2
Root	52 ± 12	0.9	33 ± 7	0.6
		Sucrose		
Leaf	30 ± 10	ND	85 ± 12	2.9
Vine	38 ± 9	ND	37 ± 9	1.3
Root	182 ± 41	ND	245 ± 18	8.4
		Starch		
Leaf	98 ± 5	1.8	291 ± 36	5.3
Vine	107 ± 18	1.9	72 ± 19	1.3
Root	1646 ± 339	29.7	615 ± 61	11.1
		Total nonstructural carbohydrates		
Leaf	139	2.6	460	9.7
Vine	154	2.8	117	2.8
Root	1,880	34	893	20.1
Plant	2,173	40	1470	33

Mean ± SD. ND, not determined.

was diverted to ethanol production (RFA, 2007). Between January 2006 and 2007, demand for bioethanol rose by over 34%, and the existing production capacity should double when facilities now under construction are brought on line in the near future (RFA, 2007). With the increasing demand for energy and the widespread distribution of kudzu across eastern North America, kudzu roots might be considered a potential new source of carbohydrate for the production of biofuels.

To determine the potential of kudzu as a biofuel, Sage et al. (2009) examined biomass and carbohydrate contents of kudzu from a location in Rockville, Maryland (near the northern edge of kudzu's distribution in North America), and from two southern locations. Below-ground biomass in Alabama exceeded 13 metric tons per hectare, and contained an average of 35% fermentable carbohydrates (sucrose, glucose, and starch; Table 10.2). Roots from Georgia of all size classes contained over 60% fermentable carbohydrates. Stems were only 1–2% carbohydrates. Biomass and carbohydrate levels in roots from Maryland were low compared to plants growing in Alabama and Georgia, being on the order of 5 metric tons of roots per hectare with 20% nonstructural carbohydrate. Based on the yield data and carbohydrate concentration, Sage et al. (2009) estimated wild kudzu stands in Alabama and Georgia could produce 5–10 metric tons of carbohydrate per hectare, which would rival carbohydrate production from corn and sugarcane fields, but without the costs associated with planting, fertilizer, and pesticides.

Kudzu is also highly responsive to carbon dioxide, although no consistent effects of rising CO_2 (up to 1,000 ppm) on root:shoot ratio have been observed (Sasek and Strain, 1988). However, kudzu, like many other potential biofuels, has a great deal of innate genetic diversity and is adapted to a wide range of environmental conditions. To be exploited, additional research would need to resolve current uncertainties, notably, how much ethanol could be produced by a hectare of kudzu roots, and whether kudzu could be economically harvested. Harvesting and processing techniques would have to be developed, and the area of harvestable kudzu would need to be assessed. Regulatory issues would also have to be resolved, particularly if kudzu were to become a renewable resource.

Overall, weeds could be a potentially important tool in adapting to energy independence and global climate change, but one with several risks and drawbacks (Raghu et al., 2006). For instance, many people are concerned that developing a large commercial market around invasive species such as kudzu would lead to constituencies interested in promoting, rather than eliminating, these species. Using harmful weeds in any large-scale revenue-generating activity would bring new controversy and political challenges.

Summary

While the harm that weeds do remains the principal focus of weed biology, we also recognize that in some circumstances, weeds may prove beneficial. As Ralph Waldo Emerson once remarked, “A weed is a plant whose virtues have not yet been discovered.” Several of these virtues have been tentatively explored in this chapter, including weeds’ potential uses as food and fuel, as medicines, as components of sustainable agroecosystems, as bioremediators, and as potential biofuels. Weeds may also have a role in adapting to climatic change, for example, as unique sources of genotypes or phenotypes that could help existing crop lines adapt to more extreme climatic conditions or to better exploit rising carbon dioxide. They might also be used to mitigate climate change, as new sources of energy, or as buffers for flooding. Additional research needs to be done to fully elucidate the positive roles weeds could play in minimizing and adapting managed plant systems to an uncertain climate; however, sufficient evidence is available to indicate a variety of potentially beneficial roles. It is important to keep in mind, however, that many of these species were purposefully introduced for good purposes, and ended up causing harm. Any new use of weeds for beneficial purposes should be undertaken with great care and restraint.

So far, we have examined weed biology in regard to function, to damage done, to management, and for benefit. In the last chapter, we will try to provide a larger context for climate change and weed science with respect to the current state of knowledge, and an assessment of current and potential threats posed by weeds and climate change, including the ecological, agricultural, and societal costs. We also explore some critical scientific questions that have not yet been addressed.

Kudzu was once considered "the miracle vine" of the twentieth century, and its virtues touted by the USDA during the 1930s for its utility as a forage crop and its ability to prevent erosion (see leaflet). Indeed, during the Great Depression, hundreds of young men were paid to plant kudzu throughout the south as part of the Civilian Conservation Corps. The problem? It grew a little too well, and by the 1960s and 1970s, its virtues had become vices, and the south was littered with "kudzuscapes" that extended as far as the eye could wander. As a consequence, kudzu remains a part of southern culture, and innovative southerners have touted its uses in everything from baskets to jelly to a potential cure for alcoholism.

Ah, you may have watched the blacksnake run.
To the shaded hole from the blistering sun,
And you may have seen the swallow's flight
And the shooting star in the deep, dark night;
But until you have watched the kudzu grow
You never have seen the fastest show.

Ollie Reeves

References

Altieri MA. 1991. How can we best use biodiversity in agroecosystems? *Outlook Agric.* **20**:15–23.

Altieri MA, Whitcomb WH. 1979. Manipulation of insect populations through seasonal disturbance of wed communities. *Prot. Ecol.* **1**:185–202.

Barbosa P, Wratten SD. 1998. The influence of plants on inveterate predators: implications to conservation biological control. In: Barbosa P (ed.) *Conservation Biological Control.* Academic Press, New York, pp. 83–100.

Blaustein RJ. 2001. Kudu's invasion into southern United States life and culture. In: McNeely JA (ed.) *The Great Reshuffling: Human Dimensions of Invasive Alien Species.* IUCN—The World Conservation Union, Cambridge, UK, 342 pp.

Bugg RL. 1992. Using cover crops to manage arthopods on truck farms. *HortScience* **27**:741–745.

Bugg RL, Ehler LE, Wilson LT. 1987. Effect of common knotweed (*Polygonum aviculare*) on abundance and efficiency of insect predators of crop pests. *Higardia* **55**:1–53.

Culley DD, Rejmankova E, Kvet J, Frye JB. 1981. Production, chemical quality and use of duckweeds (Lemnaceae) in aquaculture, waste management and animal feed. *J. World Maricult. Soc.* **12**:27–49.

Dodd JC. 2000. The role of arbuscular mycorrhizal fungi in agro- and natural ecosystems. *Outlook Agric.* **29**:55–622.

Farnsworth NR, Akerele O, Bingel AS, Soejarto DD, Guo Z. 1985. Medicinal plants in therapy. *Bul. Wolrd Health Org.* **63**:965–981.

Forseth IN, Innis AF. 2004. Kudzu (*Pueria Montana*): history, physiology, and ecology combine to make a major ecosystem threat. *Crit. Rev. Plant Sci.* **23**:401–413.

Frost RA, Wilson LM, Launchbaugh KL, Hovde EM. 2008. Seasonal change in forage value of rangeland weeds in Northern Idaho. *Invasive Plant Sci. Manage.* **1**:343–351.

Gamper H, Peter M, Jansa J, Luscher A, Hartwig UA, Leuann A. 2004. Arbuscular mycorrhizal fungi benefit from 7 years of free air CO_2 enrichment in well-fertilized grass and legume monocultures. *Global Change Biol.* **10**; 189–199.

Hatfield JL, Boote KJ, Fay P, Hahn L, Izaurralde C, Kimball BA, Mader T, Morgan JA, Ort D, Polley W, Thomson A, Wolfe D. 2008. Agriculture. In: *The Effects of Climate Change on Agriculture, Land Resources, Water Resources, and Biodiversity in the United States.* A report by the U.S. Climate Change Science Program and the Subcommittee on Global Change Research, Washington, DC, p. 21–74.

Jordan N, Vatovec C. 2004. Agroecological benefits from weeds. In: Inderjit (ed.) *Weed Biology and Management.* Kluwer Academic Publishers, The Netherlands, pp. 137–158.

Jordan NR, Zhang J, Huerd S. 2000. Arbuscular-mycorrhizal fungi: potential roles in weed management. *Weed Res.* **40**:397–410.

Karowe DN. 2007. Are legume-feeding herbivores buffered against direct effects of elevated carbon dioxide on host plants? A test with the sulfur butterfly, *Colias philodice*. *Global Change Biol.* **13**:2045–2051.

Leng RA. 1999. *Duckweed: A Tiny Aquatic Plant with High Potential for Agriculture and Environment.* Animal Production and Health Series, No. 143. FAO, Rome.

Marten GC, Andersen RN. 1975. Forage nutritive value and palatability of 12 common annual weeds. *Crop Sci.* **15**:821–827.

Marten GC, Sheaffer CC, Wyse DL. 1987. Forage nutritive value and palatability of perennial weeds. *Agron. J.* **79**:980–986.

Miller JH, Edwards B. 1983. Kudzu: where did it come from? And how can we stop it? *South. J. Appl. Forest.* **7**:165–169.

Msangi S, Sulser TB, Rosegrant MW, Valmonte-Santos R. 2007. Global scenarios for biofuels:impacts and implications. *Farm Policy J.* **4**:1–9.

Parks LJ, Tanner RD, Prokpo A. 2002. Kudzu, a valuable potential commercial resource: food, paper and chemicals. In: Keung WM (ed.) *Pueraria: The Genus Pueraria.* Taylor and Francis, New York, 290 pp.

Patriquin DG. 1986. Biological husbandry and the "nitrogen problem." In: Lopes-Real JM, Hodges RD (eds) *The Role of Microorganisms in a Sustainable Agriculture*. University of London, Wye College, pp. 81–103.

Pimental D, Patzek TW. 2005. Ethanol production using corn, switchgrass, and wood; biodiesel production using soybean and sunflower. *Nat. Res.* **14**:65–76.

Posner GH, Parker MH, Northrop J, Elias JS, Ploypradith P, Xie S, Shapiro TA. 1999. Orally active, hydrolytically stable, semisynthetic, antimalarial trioxanes in the artemisinin family." *J. Med. Chem.* **42**:300–304.

Prasad PVV, Boote KJ, Allen LH Jr., Sheehy JE, Thomas JMG. 2006. Species, ecotype and cultivar differences in spikelet fertility and harvest index of rice in response to high temperature stress. *Field Crop. Res.* **95**:398–411.

Rabatin SC, Stinner BR. 1989. The significance of vesicular–arbuscular mycorrhizal funal–soil macroinvertebrate interactions in agroecosystems. *Agric. Ecosys Environ.* **27**:195–204.

Raghu S, Anderson RC, Daehler CC, Davis AS, Wiedenmann RN, Simberloff D, Mack RN. 2006. Adding biofuels to the invasive species fire? *Science* **13**:1742–1742.

Rezvani AH, Levin ED. 2001. Cognitive effects of nicotine. *Soc. Biol. Psych.* **49**:258–267.

RFA (Renewable Fuels Association). 2007. *Industry Statistics*. Washington, DC. Renewable Fuels Association; http://www.ethanolrfa.org/industry/statistics/.

Rogers HH, Davis DE. 1972. Nutrient removal by water hyacinth. *Weed Sci.* **20**:423–428.

Sage RF, Coiner HA, Way DA, Runion GB, Prior SA, Torbert HA III, Sicher RC Jr., Ziska LH. 2009. Kudzu [*Pueraria montana* (Lour.) Merr. Var lobata]: a new source of carbohydrate for bioethanol production. *Biomass Bioenergy* **33**:57–61.

Sasek TW, Strain BR. 1988. Effects of carbon dioxide enrichment on the growth and morphology of Kudzu (*Pueraria lobata*). Weed Sci. **36**:28–36.

Schultes RE, Reis SV. (1995) *Ethnobotany: The Evolution of a Discipline*. Dioscorides Press, Portland, OR, 414 pp.

Searchinger T, Heimlich R, Houghton RA, Dong F, Elobeid A, Fabiosa J, Tokgoz S, Hayes D, Yu T-H. 2008. Use of U.S. croplands for biofuels increases greenhouse gases through emissions from land-use change. *Science* **319**:1238–1240.

Shapouri H, Duffield JA, Graboski MS. 1995. Estimating the net energy balance of corn ethanol. Agricultural Economics Report No 24 (AER 721), Economics Research Service, Washington, DC.

Sharkey TD, Loreto F. 1993. Water-stress, temperature, and light effects on the capacity for isoprene emission and photosynthesis of kudzu leaves. *Oecologica* **95**:328–333.

Shurtleff W, Aoyagi A. 1985. *The Book of Kudzu*. Avery Publishers, Wayne, NJ, 102 pp.

Stuhlfauth T, Fock HP. 1990. Effect of whole season CO_2 enrichment on the cultivation of a medicinal plant, *Digitalis lanata*. *J. Agron. Crop Sci.* **16**:168–173.

Swamy PS, Ramakrishnan PS. 1988. Ecological implications of traditional weeding regimes under slash and burn agriculture (jhum) in northeastern India. *Weed Res.* **28**:127–136.

Tanner RD, Hussain SS. 1979. Kudzu (*Pueraria lobata*) root starchas a substrate for the lysine-enriched baker's yeast and ethanol fermentation process. *Agric. Food Chem.* **27**:22–27.

Tanner RD, Hussain SS, Hamilton LA, Wolf FT. 1979. Kudzu (*Pueraria lobata*): potential agricultural and industrial resource. *Econ. Bot.* **33**:400–412.

Temme DG, Harvey RG, Fawcett RS, Young AW. 1979. Effect of annual weed control on alfalfa forage quality. *Agron. J.* **71**:51–54.

Terrill TH, Gelaye S, Mahotiere S, Amoah EA, Miller S, Windham WR. 2009. Effect of cutting date and frequency on yield and quality of kudzu in the southern United States. *Grass Forage Sci.* **58**:178–183.

Thomas L. 2006. *Bay Area Bioremediators at Work in New Orleans: Plants Prove Good at Containing Contamination Katrina Spread*. San Francisco Chronicle, April 19, 2006.

Wardle DA. 1992. A comparative assessment of factors which influence microbial biomass carbon and nitrogen levels in soil. *Biol. Rev.* **67**:321–358.

World Health Organization (WHO). 2001. The use of antimalarial drugs. Report of a WHO Informal Consultation. World Health Organization, Geneva. WHO/CDS/RBM/2001.33.

World Health Organization (WHO). 2002. Traditional medicine: growing needs and potential. *WHO Policy Perspect. Med.* **2**:1–6.

Yang YK, McEvoy JP, Wilson WH, Levin ED, Rose JE. 2003. Reliabilities and inter-correlations of reported and objective measures of smoking in patients with schizophrenia. *Schizophr. Res.* **60**:9–12.

Zandstra HB, Motooka PS. 1978. Beneficial effects of weeds in pest management—a review. *PANS* **24**:333–338.

Zimdahl, RL. 1993. *Fundamentals of Weed Science.* Academic Press, New York, NY, 454 pp.

Ziska LH, Emche SD, Johnson EL, George K, Reed DR, Sicher RC. 2005. Alterations in the production and concentration of selected alkaloids as a function of rising atmospheric carbon dioxide and air temperature: implications for ethno-pharmacology. *Global Change Biol.* **11**:1798–1807.

Ziska LH, McClung A. 2008. Differential response of cultivated and weedy (red) rice to recent and projected increases in atmospheric carbon dioxide. *Agron. J.* **100**:1259–1263.

Ziska LH, Panicker S, Wojno HL. 2008. Recent and projected increases in atmospheric carbon dioxide and the potential impacts on growth and alkaloid production in wild poppy (*Papaver setigerum* DC.) *Clim. Change* **91**:395–403.

Zobayed SMA, Saxena PK. 2004. Production of St. John's wort plants under controlled environment for maximizing biomass and secondary metabolites. *In Vitro Cell. Dev. Biol. Plant* **40**:108–114.

11 Weeds in a Time of Climate

"There is no need to sally forth. For it remains true that those things which make us human are, curiously enough, always close at hand. Resolve then, that on this very ground, with small flags waving and tinny blasts on tiny trumpets, we may meet the enemy. And not only may he be ours . . . He may be us."

Walt Kelly

The Bliss of Ignorance

Plants dominate life on earth. If one were to compare the relative biomass of the plant and animal kingdoms, plants would, by weight, make up about 90% of all living matter. Plants are autotrophs, supplying food (i.e., energy and carbon) to most life on land. Any change in plant function, particularly on a global scale, will have widespread consequences for living things.

So, how will plants respond to global anthropogenic activities? To address this question we must recognize that the production of key trace gasses (e.g., carbon dioxide, methane, nitrous oxide, etc.) will continue, because as the human population grows and societies modernize, the demand for energy and food will increase. We must also recognize that these gases, along with water vapor, absorb energy in the infrared (heat) part of the electromagnetic spectrum, and that their build-up will alter surface temperatures. The degree of increase will depend on the relative proportion of the different "greenhouse gases." For example, the amount of water vapor that the atmosphere can hold varies by more than an order of magnitude, depending on air temperature. Consequently, adding CO_2 to warm, humid equatorial climes has less of an effect on surface temperature than adding CO_2 to regions that are dry (e.g., poles and deserts). The end result, as reviewed in Chapter 2, is an increase in surface temperatures that varies across the globe (Figure 11.1), with subsequent, indirect, effects on plant biology through changes in temperature or accompanying changes in precipitation.

While we recognize warming as having an obvious effect on plant function, we must also acknowledge that the rise of carbon dioxide, the principal greenhouse gas, will have a direct influence on plant biology. Plants require four abiotic resources (sunlight, nutrients, water, and carbon dioxide) to complete their life cycle. Any change in the availability of these resources will result in a change in plant and ecosystem function. For example, suppose that light, or water, or nutrient availability had increased by 20% since 1970, would it be safe to

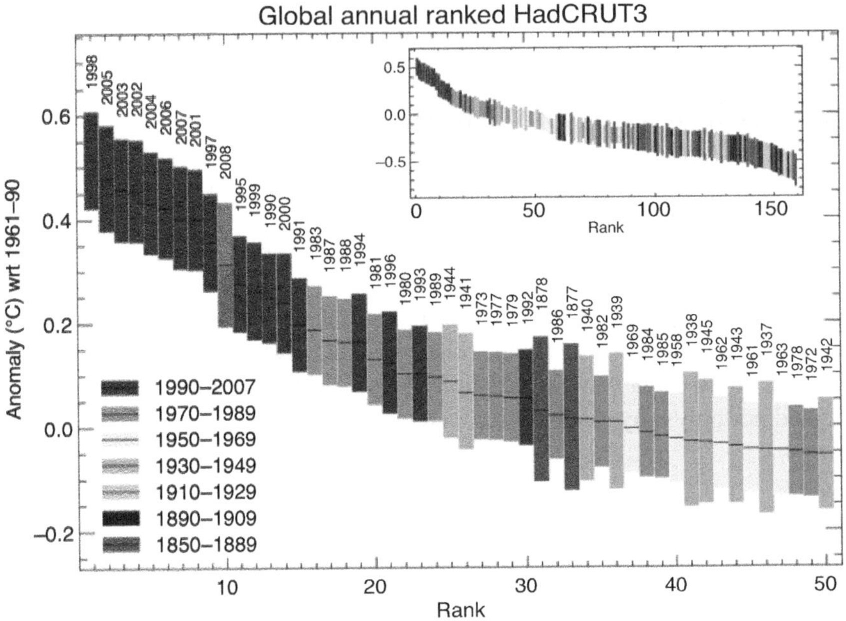

Figure 11.1 Climate scientists at the Met Office Hadley Centre and the Climatic Research Unit (CRU) at University of East Anglia maintain the global climate record for the World Meteorological Organization (WMO). The global mean temperature for 2008 was 14.3°C, making it the tenth warmest year on a record that dates back to 1850. Updated version available in color online at: http://www.metoffice.gov.uk/climatechange/policymakers/policy/slowdown.html.

say that plants (including weeds) would be both impacted by, and respond differentially to, such a change? As plant biologists, would not we want to understand such a change in the context of plant function, from the gene to the ecosystem?

Since 1970, the amount of carbon dioxide in the atmosphere has, in fact, risen approximately 20% (327–389 ppm from March 1970 to March 2009, www.cdiac.gov). Clearly, there are and will be global differences in how plants respond to this very sudden change in a necessary resource.

In spite of the obvious ties (direct and indirect) between the build-up of greenhouse gas emissions and plant biology, assessing the impact of rising carbon dioxide and climate change on plant function remains under-appreciated. For example, from the latest IPCC (Inter-governmental Panel on Climate Change) fourth assessment report (2007), only 2 of the 20 chapters deal with plant biology, and only indirectly as either ecosystem services or food and fiber. Similarly, the recent "Global Climate Change Impacts in the United States" (Karl et al., 2009), by the U.S. Global Change Research Program has less than three pages on weeds and other pests.

Too often anthropogenic climate change is treated in simplistic terms (e.g., "*51 Things You Can Do to Make a Difference*," Time magazine, April 2, 2007); or, as a biological stimulus ("*We are living in an increasingly lush environment of plants and animals as a result of the carbon dioxide increase. This is a wonderful and unexpected gift from the industrial revolution,*" Robinson and Robinson, 1997).

Clearly, how one views the concept of global climate or global warming can be strikingly different (e.g., bane or boon). As weed scientists, however, we emphasize the concept of *change*. That is, whether we are examining temperature, or precipitation or carbon dioxide, we know that the growth and fecundity of plant species will demonstrate both intra- and interspecific variability to any shift in an environmental parameter. Certainly, some changes can benefit people, particularly with respect to exploiting carbon dioxide responsiveness among existing crop cultivars as a means to boost yields. For example, significant intraspecific variation in yield response with a doubling of current CO_2 levels has been observed in cowpea (Ahmed et al., 1993), rice (Moya et al., 1998; Ziska et al., 1996), soybean (Ziska and Bunce, 2000), and wheat (Mandersheid and Weigel, 1997; Ziska et al., 2004), inter alia.

But it should also be emphasized that climate change and carbon dioxide levels do not distinguish between those plant species that are anthropogenically beneficial, and those that do harm. Weeds are defined, in general, as those plant species that do harm, whether at the agricultural level (managed systems), in the environment (unmanaged systems) or for public health. As we have tried to illustrate throughout this book, many of the anticipated climatic changes indicate a destabilizing set of environmental conditions that are likely to favor those characteristics associated with weed biology and fecundity. Weeds are, in fact, inherently adapted, and supremely malleable to environmental disturbance and transition. As such, they are poised to be the ultimate beneficiaries of global climate change.

What Are the Implications?

To understand the implications is to understand the nature of individual versus group change. That is, by altering the environment (e.g., temperature) or increasing a resource (e.g., CO_2), we not only change the growth of an individual, but also those interactions between genotype and environment that alter the growth of the entire plant community. As a result, we perceive many implications for weed science.

Managed Systems

Agriculture, in its simplest arrangement, can be characterized as a managed plant community that is composed of a desired plant species (the crop) and many undesired plant species (the weeds). As has been acknowledged since the inception of agriculture, weeds represent the largest limitation in regard to plant yield. For example, in China, 10 million metric tons of rice is estimated to be lost annually due to weed competition (Zhang, 2001). This is sufficient rice to feed 56 million people for 1 year (Labrada, 2007). Yield losses in rice can approach 100% if weeds are not controlled (Ampong-Nyarko and De Datta, 1991).

Crops and weeds will respond differently to environmental change. In the case of rising carbon dioxide, it is certainly possible, based on photosynthetic pathway, that a C_3 crop could be favored over a C_4 weed. However, as we have seen, a number of the studies conducted to date suggest that for plant species within an agricultural system, weeds, rather than crops, are likely to show the largest relative response to carbon dioxide and/or climate. That is not to say that individual crop species cannot show a positive response, but the greater response of weedy species may result in increased competition and exacerbated yield losses. While photosynthetic biochemistry may distinguish CO_2 responses in some cases, it is clear that a simple C_3 crop, C_4 weed explanation is inadequate to address the results that have been observed to date. A state by state assessment conducted by Bridges (1992) for crops in the United States found that, on average, each crop competes with 8–10 weeds that are considered "troublesome." These weeds are both annuals and perennials, C_3 and C_4. Many of them are simply wild plants of the same genus or species (e.g., rice and red rice, oat and wild oat, sorghum and shatter cane, potatoes and nightshades).

Weed populations in agricultural systems have the advantage of diversity: they consist of many different species, each with a full array of genotypes. In contrast, breeders develop crops to have high genetic and phenotypic uniformity, but not necessarily CO_2 responsiveness (Ziska and Blumenthal, 2007). While this selection is advantageous in modern agriculture, with greater crop uniformity of obvious benefit as farm size increases, reliance on a narrow genetic base may confer a disadvantage on current crop lines in regard to weed–crop competition if the environment is, in fact, nonuniform. That is, the greater competitive ability of weeds to respond to CO_2 or climate change may be related to their greater genetic diversity relative to the crop monoculture (Treharne, 1989). Whether modern crop breeding can change tactics, that is, begin to select for genetically diverse crop lines that are well adapted and competitive in a changing climate, or that stay optimized to the current CO_2 concentration, represents an obvious challenge for agricultural scientists in the twenty-first century (Newton and Edwards, 2007).

Native Systems

For relatively unmanaged plant communities, invasive plant species (IPS) represent an obvious threat to community function. Millions of acres of productive rangelands, forests, and riparian areas have been impacted by invasives with a subsequent loss in native flora. Estimates are that more than 200 million acres of natural habitat, principally in the western United States, have been lost to invasive, noxious weeds, with an ongoing loss of 3,000 acres per day (Westbrooks, 1998).

As we have discussed (see Chapter 6), responses of invasive plants to the changing climate and atmosphere will affect native species in natural or less managed ecosystems. For example, the increase in carbon dioxide could benefit cheatgrass, increasing the frequency and spread of natural and anthropogenic fires in the intermountain West (Ziska et al., 2005a). Some data suggest that, on average, invasive species may show a stronger response than noninvasive species to both recent and projected changes in carbon dioxide relative to other plant species (Ziska and George, 2004). However, other analyses suggest that there is little difference in response among these categories (Dukes, 2000). A handful of field-based studies suggest that rising levels of carbon dioxide preferentially increase the growth of some invasive plant

Table 11.1 A summary of community-level experiments to determine if rising CO_2 favors invasive species within plant communities.

Species	Community	Favored?	Reference
Yellow starthistle	California serpentine grassland	No	Dukes, 2002
Yellow starthistle	California valley grassland	Yes	Dukes et al., unpublished
Honey mesquite	Texas prairie	Yes	Polley et al., 1994
Japanese honeysuckle	Forest under-story	Yes	Belote et al., 2003
Microstegium vimineum	Forest under-story	No	Belote et al., 2003
Cherry laurel	Forest under-story	Yes	Hattenschwiler & Korner, 2003
Red Brome	Desert	Yes (some years)	Smith et al., 2000

species within some plant communities (Table 11.1), although results are not uniform. As with agronomic weeds, the basis for the greater relative response to atmospheric carbon dioxide, where observed, is unclear. The greater response of invasive weeds to rising CO_2 is consistent with the hypothesis of Blumenthal (2005, 2006) that fast-growing weedy species that benefit more from increases in resource availability will also benefit most from escaping their natural enemies.

An additional complication is that of climate, particularly temperature. Bradley et al. (2009) used bioclimatic envelope modeling to show that climate change could result in both range expansion and contraction for five widespread and dominant invasive plants in the western United States. They projected that yellow starthistle and tamarisk (*Tamarix* spp.) are likely to expand with climate change, whereas cheatgrass and spotted knapweed are likely to shift northward, leading to both expansion and contraction. Hence, while some invasive species could become an additional problem with climate change, it is possible that the retreat of some invasive species could create restoration opportunities as well.

Societal Implications

By impacting plant communities, climate and carbon dioxide will alter the goods and services provided by these communities. As discussed in Chapter 7, there are also likely to be effects on human health. While the implications of climate and atmospheric change for weeds and their effects on health are not entirely elucidated, it is clear that there are a number of likely interactions.

For example, there are over four dozen known allergenic weed species, and at least one, common ragweed, which shows a strong response to rising CO_2 in both laboratory and field studies (Wayne et al., 2002; Ziska et al., 2003). There are over a hundred different weed species associated with contact dermatitis, an immune-mediated skin inflammation. Among these, poison ivy appears to show a strong quantitative and qualitative response to CO_2 (Mohan et al., 2006). There are over 700 weed species that are poisonous to humans. Ingestion of plant material continues to be a very common exposure for humans (particularly children) and can account for nearly 100,000 calls to national poison centers annually (Watson et al.,

2004). Pediatric patients comprise more than 80% of plant-related exposures. Recent data for cassava suggest that rising CO_2 can increase the concentration of cyanotic glycosides and increase toxicity (Gleadow et al., 2009). A number of common agronomic weed species have also been used in pharmacology. Preliminary results indicate that both temperature and carbon dioxide are likely to influence the production and concentration of these secondary compounds (e.g., atropine and scopolamine from Jimsonweed; Ziska et al., 2005b). There may be additional implications with respect to the spread of human disease. Weeds do not, of course, act as disease vectors. However, animal vectors, notably rodents and mosquitoes, rely on plants, potentially including agronomic weeds and IPS, as a principal food source. In turn, weed growth, pollen and seed production are likely to change as a function of CO_2 and temperature. For example, pollen on open ponds can serve as food for mosquito larvae (Ye-Ebiyo et al., 2000). Alterations in the distribution of plants providing food and habitat for vectors (e.g., rodents) remains an obvious, but underappreciated factor in disease vector transmission.

Implications: Weeds and Beyond

The changing atmosphere and climate will affect weed biology, with many consequences for society. In addition to weeds, other pests of human consequence, including insects and pathogens, will respond to the changing conditions. The breadth and extent of these impacts is beyond the scope of this book (for a more comprehensive assessment, see Dukes, 2000; Dukes et al., 2009; Scherm and Coakley, 2003; Sutherst et al., 2007); however, we hope to offer here, in some small measure, a brief overview of these interactions, particularly for invasive insects and pathogens, in the hope that they will provide a tentative framework for additional inquiry.

Pathogens

Plant pathogens are widely considered among the most important invasive pests for plants and livestock. Between 1940 and 1970, less than five nonindigenous plant pathogens arrived into the United States per decade (National Research Council, 2002); however, by the beginning of the 1990s, over 230 exotic plant pathogens had become established in the United States, including a number of significant agricultural pests (Table 11.2)

There are a number of potential climate change drivers that are likely to exacerbate the spread and success of invasive pathogens. These include: (A) increased weather severity, for example, it is thought that fungal spores of *Phakopsora pachrhizi* (Asian soybean rust) arrived from South America carried by winds associated with hurricanes Francis and Ivan in 2004; (B) water and precipitation, in addition to wind, urediniospores and aeciospores (e.g., stem rust) germinate when in contact with free water; (C) precipitation is needed for effective spore deposition and for regional spore transport and temperature. As with moisture, temperature is a key environmental determinant of growth and sporulation of a number of invasive pathogens. Warmer winters projected with climate change would remove

Table 11.2 Examples of nonindigenous plant pathogens of agriculture introduced to, or first detected in, the United States since 1990.

Pathogen	Disease	State or region	Reference
Claviceps africana	Sorghum ergot	Southern/central states	*Plant Disease* 82:356 (1998)
Phakopsora pachyrhizi	Soybean rust	Southern states	*Plant Disease* 89:774 (2005)
Tilletia indica	Karnal bunt	Southwestern States	*Plant Disease* 81:1370 (1997)
Cucurbit aphid-borne yellow virus	Cucurbit yellows	California	*Plant Disease* 77:1169 (1993)
Oidium sp.	Powdery mildew of tomato	California	*Plant Disease* 80:1303 (1996)
Potato mop-top virus	Triparte pomovirus	Maine	*Plant Disease* 87:872 (2003)
Phytophthora porri	Cabbage rot	Wisconsin	*Plant Disease* 78:1123 (1994)
Plum pox virus	Sharka	New York, Pennsylvania	*Plant Disease* 84:202 (2000)
Xanthomonas axonopodis Pv. Citri	Citrus canker	Florida	*Plant Disease* 85:340 (2001)

Updated from Scherm and Coakley (2003).

this thermal limitation for pathogen morbidity with a subsequent increase in the potential range and distribution of invasive plant pathogens (Burden and Elmquist, 1996).

Insects

Invasive insect pests are recognized global vectors of agricultural and environmental damage. As with pathogens, increased introductions of invasive insects are likely with global trade, with climate change contributing an additional factor in their establishment and spread. Important aspects of climate change include: (A) increasing frequency of weather extremes, and particularly high winds, which can distribute insects over long distances (e.g., Desert locusts; Rosenberg and Burt, 1999); (B) changes in precipitation regimes and soil moisture, which can alter survival rates. For example, the introduction and spread of Swede midge (*Contarinia nasturtii*), an invasive pest of cruciferous crops in New York, may be related to above average precipitation during the time of its introduction (Olfert et al., 2006). Precipitation extremes (droughts and floods) are associated with several changes in insect behavior and fecundity (Watt and Leather, 1986); (C) temperature is among the central regulating factors controlling the development and fecundity of invasive insects, consequently, climate warming is likely to increase their geographical distribution (both in latitude and altitude) and/or enhance their capacity for over-wintering (Fleming and Tatchell, 1995).

Interactions: Weeds and Insects and Pathogens

As illustrated in Chapter 5, weeds can serve as reservoirs for plant pathogens that impact crops. As such, weeds can act as the initial inoculum for epidemics in crops and can exacerbate

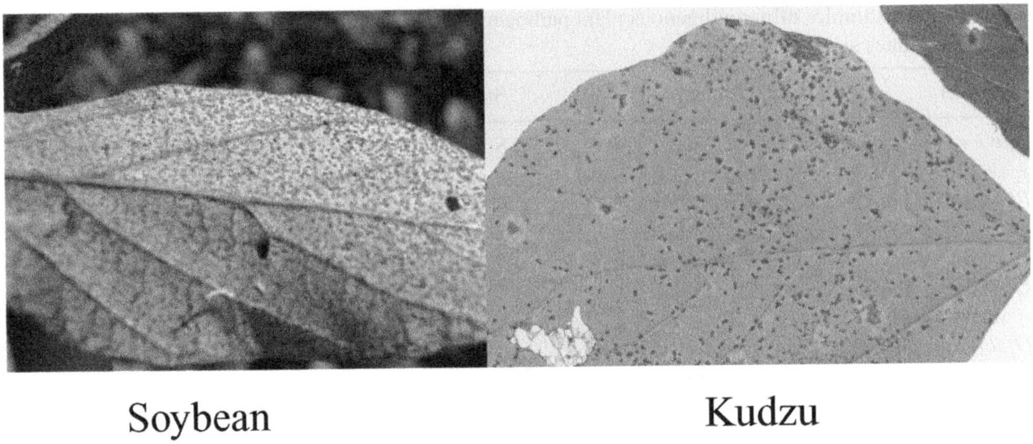

Figure 11.2 Kudzu, an invasive weed as an alternative host for Asian soybean rust (*Phakopsora pachyrhizi*).

existing problems by providing multiple hosts for reproduction. American black nightshade, a serious weed problem in several solanaceous crops, is a host for multiple pathogens (e.g., Tomato spotted wilt virus) that can impact these crops. Weeds may also serve as reservoirs for insect pests. A study of cassava root scale (*Protortonia navesi*), a recent pest in the Brazilian "Cerrado" that causes qualitative and quantitative damage by sucking plant sap, identified 15 weed species as hosts for the insect (Oliveira and Fontes, 2008). Weeds can also serve as hosts for insects that spread disease. For example, Russian thistle can act as a host for beet leafhopper (*Circulifer tenellus*), a carrier of the curly top virus (Curtovirus spp.). Finally, weeds can provide habitat for insects that provide a level of natural biological control of other insect pests.

Clearly, changes in CO_2 and in climate, particularly temperature and precipitation, can be expected to ilicit a complex series of interactions that will affect the relationship between pathogens, insects, and weeds as hosts. Kudzu, as already discussed, is a ubiquitous invasive weed, distributed throughout the southeastern United States and a host for Asian soybean rust (*Phakopsora pachyrhizi*), a significant threat to soybean production (see Figure 11.2). Soybean yield reductions and production cost increases have been attributed to rust in Africa, Asia, Australia, and South America. By 2009, rust had been confirmed in over 270 counties in 15 states (http://sbr.ipmpipe.org/cgi-bin/sbr/public.cgi).

Obviously, given its potential as a reservoir for the pathogen, climate-induced migratory changes in kudzu are of particular concern. One of the most interesting forecasts in that regard was made almost two decades ago for kudzu and global warming. Tom Sasek and Boyd Strain at Duke University observed that the latitudinal distribution of kudzu at that time was limited to regions south of the Mason-Dixon line by low winter temperatures of $-15°C$ (see Figure 7 in Sasek and Strain, 1990) but projected that with global warming, particularly mild winters, that kudzu would increase its northward migration. A more recent evaluation of current kudzu distribution indicates a significant shift in the northern boundary of kudzu

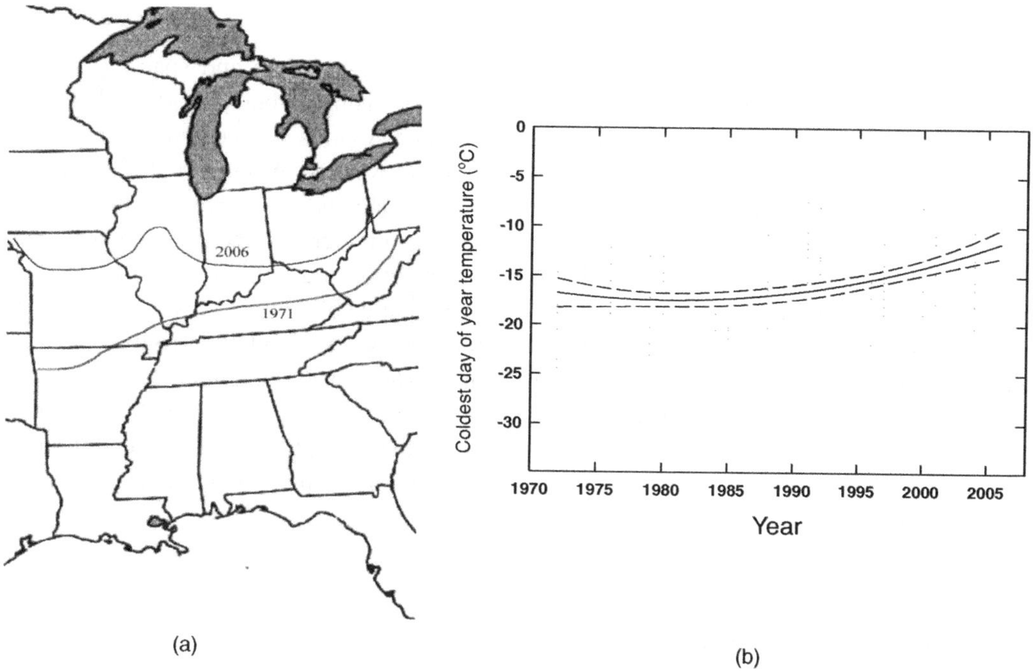

Figure 11.3 **(A)** Shift in northernmost detection of kudzu (*Pueraria lobata*) from 1971 to 2006. The 1971 line is from Clyde Reed, "Common Weeds of the United States" a USDA-ARS publication. Estimates of kudzu distribution in 2006 for Nebraska, Missouri, Illinois, Indiana, Ohio, and Pennsylvania were evaluated using three separate sources: (a) National Resource Conservation Service NRCS), database of invasive U.S. species (plants.usda.gov/java/profile?symbol = PUMO); (b) the National Agricultural Pest Information Service (NAPIS) in cooperative agreement between the Animal Plant Health Information Service (APHIS) and Purdue University as part of their Cooperative Agricultural Pest Survey (CAPS) program (ceris.purdue.edu/napis/pests/weeds/imap/kudzu.html); and (c) the Department of Natural Resources (DNR) for the states of Pennsylvania, Ohio, Indiana, Illinois, and Missouri, including the publication of "The Green Plague Moves North" by the Illinois DNR. At least one county was chosen as the northernmost county within a state reporting a new kudzu population in order to determine the northernmost range. Counties within a given state were only included if the presence of kudzu was verifiable from at least two of these sources. Southeastern Nebraska and Western Pennsylvania were considered the end points of the Midwestern populations. Specific counties sampled were Otoe Co (NE), Reynolds, Phelps and Howard Co (MO), Fayette, Shelby, Macon and Woodford Co (IL), Vanderburgh, Martin, Owen, and Johnson Co. (IN), Belmont, Athens, Scioto Co. (OH), Allegheny CO. (PA). **(B)** Average temperature ($\pm 95\%$ confidence interval) for the coldest recorded day of year for 16 counties along northern limit for kudzu (Figure 11.4a) in the Midwestern United States (1972–2006). Meteorological data for a given county was evaluated with respect to long-term multi-year data availability, and proximity to urban heat islands. County meteorological data were evaluated from a larger database contained within the National Climatic Data Center (NCDC) in Asheville, NC (hurricane.ncdc.noaa.gov), from 1972 through 2006. An average temperature of $-15°C$ has been suggested as the lower temperature limit for over-wintering in kudzu (Sasek and Strain, 1990).

that appears related to an increase in the minimum winter temperatures (see Figure 11.3). How much of the latitudinal shift is related to increasing winter temperatures and thermal limits of kudzu is still unclear, but the northward spread is consistent with the projections of Sasek and Strain (1990). Their study remains unique in projecting the interactions between climate change, weeds, and pathogens.

Weed Management

As we have seen, every aspect of weed biology that varies in time and space, from the genome to the plant community, is likely to be affected by rising carbon dioxide and climate change. The extent that these changes will impact human systems will be determined, in no small measure, by the rate and success of adaptations in weed management strategies (see Chapters 8 and 9). In examining the likely impact of climate change on weed management, it is important to remember that past is not prologue. The number of weeds that are herbicide resistant within developed countries has increased in an almost logarithmic fashion in the last decade (www.weedscience.com). To assume that the practice of applying a single herbicide to a single crop will continue unabated even if climate change were not occurring is specious indeed. Climate and CO_2 add management uncertainties that vary from changes in the range, habitat, and abundance of a given weed, to changes in the efficacy of chemical, biological, physical, and cultural weed control. It seems unlikely that a simple, single means of weed control (e.g., glyphosate) will continue to be a viable option for many land managers by the end of the twenty-first century.

Uncertainties and Limitations

There are three broad areas of uncertainty in relation to weeds and climate change: (A) what are the biological characteristics related to function, fecundity, and habitat that are likely to determine biological "winners" and "losers" within the weed community; (B) how will these biological changes increase or decrease the ability of weeds to inflict damage on plant and human systems; and (C) what are the implications for weed management and control?

Biology

Many of the biological uncertainties regarding the impact of climate and rising CO_2 arise from the experimental data, much of which reports the impact of single environmental factors (e.g., CO_2) on weed biology. In addition, these data tend to focus on those parts of the plant easy to measure, for example, leaves, plant growth, flowering times, etc., and are mostly from plants grown in controlled environments and/or without competition. However, to adequately assess the utility of experimental data on the response of weeds to climate change it will be necessary to "scale-up" or "scale-down" from the whole-plant level to different organizational levels involving multi-variable interactions. Which organizational levels require further evaluation? There are a number of studies that have reported on the response of individual weeds to a "doubling" of CO_2 concentrations (e.g., Patterson, 1995), but much less is known regarding the impact of rising CO_2 in combination with other anticipated climatic changes with respect to spatial or temporal extremes. For example, we know little about how climate may alter phenotypic expression among weed biotypes; at the ecosystem level, we know little about how temperature, precipitation, and carbon dioxide are likely to influence the range and composition of agronomic or invasive weeds.

Damage

Understanding how weeds will respond to a rapidly changing climate is fundamental to assessing their likely impact. Hence, uncertainties at the biological level become magnified when quantifying weed-induced damage in human and natural systems. Two challenges in this regard are: (1) Estimating damage from weeds in an environmentally changed environment. How will changes in temperature and precipitation alter the growth, distribution, and impacts of weed species? (2) Identifying biological characteristics that confer a competitive advantage in a changing climate and atmosphere. Identifying the traits associated with superior performance in future climate/CO_2 conditions is a major challenge, as responsiveness to these conditions is likely to depend on a myriad of different biological parameters (e.g., seed rain, germination ability, leaf development, allelopathy, etc.). Yet, developing this understanding is necessary if we are to reliably forecast the potential damage of a given weed within a plant community.

Management

Management and mitigation decisions depend on evaluations of the potential damage associated with the presence of a given weed, and where and when the weed might grow. Changing climate and CO_2 will alter distributions and impacts of weeds, but also the efficacy of a number of current weed management strategies, including herbicide application. No comprehensive assessment of how management strategies will be affected is currently available. There are a few reports of changing herbicide efficacy (e.g., Archambault, 2007), but the biological basis for these changes has not yet been entirely elucidated. In addition, environmental factors such as temperature, precipitation, wind, humidity, and carbon dioxide will also influence herbicide efficacy.

In spite of the recognition that CO_2 and climate change can fundamentally alter weed biology, little attention has been given to the weed management aspect of global change/global warming. Yet, given the potential threat posed by rapid anthropogenic change, it is clear that the testing and adoption of new management techniques will be essential in order to minimize the already considerable impact of weeds on managed and native systems.

Future Direction

As with any initial integration and assessment, the observations given here are not meant to be all-encompassing, but rather to illustrate the scope of potential interactions between weeds, carbon dioxide, and climate. Weed science is just beginning to consider climate change, and there remain many fertile areas of inquiry. We hope that this book encourages anyone interested in pursuing research in this field to press forward. We recommend that, where possible, future research designs incorporate many simultaneous changes in environmental parameters.

Threat Recognition

As the rapidly growing human population strains the global food supply, scientists recognize that weeds threaten food security. As invasive weeds proliferate in natural habitats, environmentalists recognize the threat they pose to ecosystem function and biodiversity. As weeds spread through urban areas, public health workers recognize that the pollen of some weeds provokes allergic reactions. The cost of all these threats, in economic and ecological terms, is high. In U.S. agriculture, weeds cause an estimated reduction of 12% in crop yields under best management practices (i.e., without herbicide use; Bridges, 1992), causing losses of about $33 billion annually, based on a total U.S. crop potential value of more than $267 billion per year (USBC, 1998). Pimentel et al. (2005) suggest weeds cause $34.6 billion per year in control costs and economic losses in managed systems alone. These estimates do not include indirect costs of weed biology; for example, the cost of land taken out of development by invasives, the health care costs related to plant-based allergens, etc. Although there are benefits from some weeds (e.g., reduced soil erosion), the net cost in economic and environmental terms is unquestionably substantial.

Clearly, weed impacts need a more complete accounting and the costs are likely to change as weeds (and their competitors, herbivores, and pathogens) respond to the changing environment. Efforts to assess the biological impacts of CO_2 and climate change have, in our opinion, dedicated too few resources to studying how environmental changes will alter the economic impacts of weeds. We strongly recommend that policy makers, scientists, NGOs, and others involved in climate change synthesis and assessment consider weeds in any future evaluations. While the net impacts of weeds may well increase, targeted biological research would be needed to provide enough data for a robust economic analysis.

Research Priorities

As the field moves forward, we see a need for additional experimental data in four key areas: (a) weed/crop interactions; (b) invasive plants in relatively unmanaged systems; (c) weeds and public health; and (d) weed management.

Several key unknowns regarding weed/crop interactions require additional information. More field research is needed to quantify competition between weeds and perennial crops; additional information is needed for crop–weed interactions for species with the same photosynthetic pathway (e.g., C_3 versus C_3 crop:weed); in particular, future research should identify how environmental changes (and climate change in particular) affect these competitive outcomes. If researchers are to quantify the basis for determining “winners” and “losers” in a changed environment, they must assess how species differ in responses to environmental change, particularly in terms of traits such as seed dormancy, germination and emergence, allometric and phenological development, carbon allocation and allelopathy. Preferably, to achieve such a goal, research would take place outdoors, and involve realistic mixes of species.

There is evidence to suggest that, in many cases, CO_2 and temperature could enhance invasive species success (e.g., Song et al., 2009). If so, how do climate and/or CO_2 differentiate invasive weeds from native vegetation? Are the same biological responses observed for weed–crop competition sufficient to explain greater competitive response among IPS? How will IPS range and distribution respond to warming winters, or changed precipitation events? Will climate differentially affect IPS and predators, resulting in "new" invasives, that is, agronomic weeds that may become IPS if they "out-migrate" their pathogens and herbivores?

From the initial studies described in Chapter 7, it is evident that plant responses to climate and atmospheric change will affect public health. But many questions remain. What other weed species are likely to increase allergenic pollen production in response to changes in climate and CO_2? What are the implications for allergies and asthma? Will cases of contact dermatitis become more common, and more severe? Can we expect toxicological changes in poisonous plants? How will CO_2-induced changes in food quality affect human nutrition and health, or animal nutrition and health? Given that a number of weedy species are sources of medicine, how will climate and CO_2 alter their distribution or the production and concentration of secondary plant compounds? As weed distributions change with CO_2 and climate change, will the ability of mosquitoes or rodents to spread disease be affected?

There are a number of key research priorities related to management: How commonly will herbicide efficacy decrease (or increase), and what will be the implications for herbicide levels in the environment? How do environmental changes alter herbicide resistance? What are the cultural, biological, and physical aspects of current weed management practices most vulnerable to CO_2 and climate? Understanding the vulnerability of existing management systems is essential in order to identify measures that can strengthen weed control and lessen damage in the future. What approaches can be adapted to increase herbicide effectiveness? If chemical control is less efficacious, how can land managers compensate? How are other types of weed control likely to be affected? For example, if climate/CO_2 increases root to shoot ratio in perennial weeds (e.g., Canada thistle; Ziska et al., 2004), tillage efficacy is also likely to be reduced. Will biological control agents and weeds respond differently to climate and CO_2, and if so, how will this alter predator–prey relationships?

Answers to these research questions will also bring up an opportunity for reevaluation of current management practices: Should greater attention be paid to restoration of native habitats around agricultural fields to prevent invasion? Are there cultural measures (e.g., washing of all vehicles to prevent seed movement) that should be mandated around agricultural fields? Should integrated pest management (IPM) strategies be modified to reflect changing environmental conditions?

Models and Forecasting

Science is reductionist in nature, global environmental changes are not. Given the unprecedented scale and scope of anticipated changes in the atmosphere and climate, it is tempting to ignore the large number of uncertainties as being unquantifiable, and to focus on one or two parameters (e.g., CO_2 in isolation from other changes) that can be well defined. In general, such an approach is consistent with that of many experimentalists who work in large part on quantifying one or two variables, usually in the short term (1–2 years). Such a reductionist approach is certainly valid, as it provides key information, and insight to mechanisms. However, this approach has limited utility in predicting broad consequences from global climate change. We see a need for an approach that integrates information from experiments and observations of weeds to model species dynamics and demographics across geographically relevant scales. Such an approach could also facilitate the incorporation of new information on weed responses into an improved knowledge-based systems approach to weed management.

Another aspect of weed management, forecasting, will become increasingly important. There will be no simple, universal fix to weed problems that crop up in the wake of environmental changes. However, improved forecasting, particularly of the most pernicious weeds, may provide an opportunity to limit damage. Despite the threats posed by IPS, few long-term field observations have documented distribution and abundance of IPS as a function of environment. Such monitoring could help identify communities and climates most vulnerable to invasion (Lawton, 2000). For example, in Australia, efforts are underway to integrate invasive plants and climate change into a "threat index" as a means to identify areas of vulnerability (Crossman et al., 2008). Similar efforts are underway on the part of the Victorian government in Australia to examine climate change and weed distribution (DPI, 2008). In the United States, long-term ecological research (LTER) sites, originally designed to monitor ecosystems of interest for conservation, have also compiled preliminary data on community composition, invasive species, and climate change (Roy et al., 2004). Assigning additional responsibility for monitoring climate and invasives to existing LTER sites is one strategy to increase baseline biodiversity information at minimum cost. An expansion of a variety of monitoring efforts could be invaluable with respect to forecasting. For example, some transportation corridors are known to facilitate dispersal of invasive species (e.g., Gelbard and Belnap, 2003; Tyser and Worley, 1992). Placement of monitoring stations along key transportation corridors, particularly near large ports, could help identify new and rapidly spreading invasive threats. Information about the phenology, distribution, and response traits of a given invasive species would greatly facilitate forecasting, and would help managers devise suitable containment or eradication strategies for that species. Developing a monitoring network to conduct observations and collect data would be an essential first step in forecasting and mitigation of climate- and CO_2-induced impacts regarding the range and spread of destructive weed species.

Communication/Relevance

Lastly, while many scientists now recognize the implications of climate change for weed biology, the general public, and policy makers are likely to overlook the economic and environmental implications until there is a call for action from those studying the problem. Ideally, scientists should provide real-world socioeconomic examples that consumers and policy makers relate to. Researchers can provide specific examples of weed impacts on food security, invasive plants' impacts on biodiversity, and so on. While the science may be inherently worthy in its own right, it will be seen by the public as an "academic exercise" (at best) unless it can be related to real-world problems. Scientists will need to communicate and share data among themselves, but will also need to communicate to larger audiences via various types of media if they are to garner the support and subsequent resources needed to address these issues.

Such support is needed. The scientific uncertainties highlighted throughout this book remain unknown, in part, because the scientific and fiscal resources needed to address them have not been available. Funding has been made available to study invasive species, (usually those that are viewed as threats to U.S. Homeland Security, which may or may not include weeds), but the overall subject of weed biology and anthropogenic environmental change has rarely been specifically addressed. Government agencies have funded individual projects on this general topic, but with a species-by-species, location-by-location emphasis. Such an approach is fragmented and solitary. Public and policy makers need to fund coordinated research that addresses the most important scientific uncertainties while beginning to concentrate on the recommendations needed to reduce uncertainty with respect to impact. Science that can lead to unifying principles is indispensable to long-term, sustainable solutions.

A Word in Your Ear

As of this writing, polls suggest that the majority of Americans do not believe that anthropogenic climate change will affect their lives. According to a Gallup poll conducted in March of 2009, only 38% of Americans, similar to the 40% found in 2008, believe it will pose "a serious threat" to themselves or their behavior. In spite of the overwhelming scientific consensus, the issue has unfortunately become one of political polemic and not scientific inquiry. Even among those who "believe" in the science, the idea that mitigation of carbon dioxide could stop or slow global warming seems to be prevalent; yet, until the recent global financial crisis, global output of carbon dioxide exceeded even the "business-as-usual" scenario of the IPCC (Figure 11.4; Canadell et al., 2007; Raupach et al., 2007).

Psychologically, it is understandable, given the global nature of climate change (and the implications for lifestyle change) that people are tempted to ignore or deny the underlying science (Stoll-Kleemann et al., 2001). Even among those who accept the science, there remains hope that people will come together to mitigate and prevent drastic climatic changes. Yet, global climate change has already extended growing seasons, changed species distributions, and altered the phenology, breeding, and migration of flowers (Fitter and Fitter, 2002), insects (Parmesan et al., 1999), birds (Cotton, 2003), and mammals (Walther et al., 2002), with an overall advancement of spring of 2.8 days in the Northern hemisphere (Parmesan, 2007). It would be very odd indeed if weed biology remained unaffected—and, of course, as we have shown throughout this book, it is not.

Ultimately, as global environmental changes cause a massive shift in Earth's species and biological processes, the responses of weeds will make up only a small portion of this shift—although an underappreciated portion, in our opinion. Unfortunately, the threat now

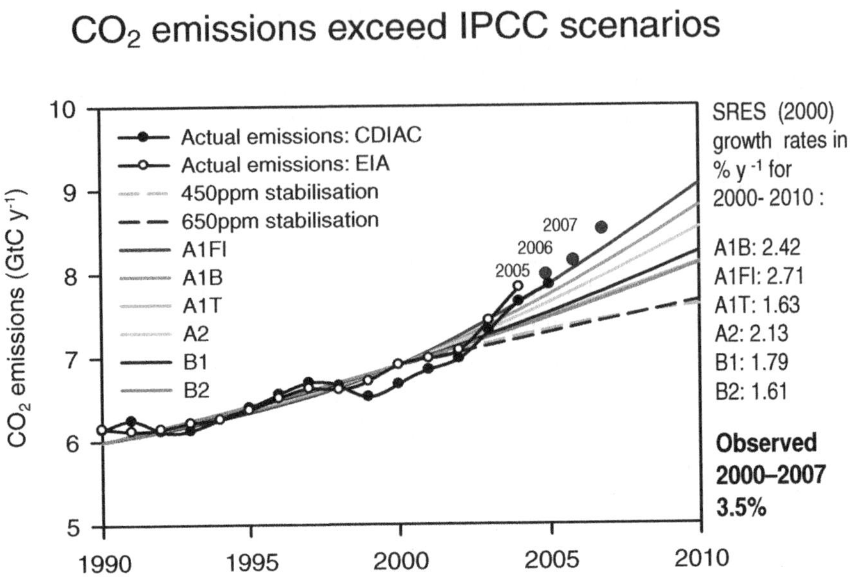

Figure 11.4 Different IPCC (2007) scenarios for anticipated changes in global CO_2 emissions (in gigatons per year). Actual CO_2 emissions are shown by the black dots. Through 2007, actual CO_2 emissions exceeded all IPCC projections, 3.5% versus estimates of 1.6–2.4%. Emissions declined somewhat following the recent global financial crisis. (Modified from Canadell et al., 2007; Raupach et al., 2007. Updated, color version available online at: http://www.globalcarbonproject.org/carbonbudget/index.htm)

posed by weeds to food security, environmental stability, and human health remains largely ignored. For decades, weed science has been extraordinarily successful in synthesizing a broad base of interdisciplinary knowledge to devise cheap, safe, and effective methodologies to manage weed threats. Weed scientists may simply be victims of their own success.

In researching and synthesizing the material for this book, it became very clear that a crucial scientific challenge currently facing weed science is that of climate change and rising atmospheric CO_2. While there may be a role for weed science in mitigation of climate change (e.g., use of herbicides in no-till systems); we emphasize the need for adaptation of weed science and management practices to climate change. How can this be accomplished? Weed science, like most human activities, projects the future by extrapolating from the past. Although there is still much to be known, we believe it is unwise to simply extrapolate past patterns into the future in an uncertain climate. Hence, there is an unprecedented scientific need for ecological assessment, threat recognition, and effective management strategies. While these are formidable challenges, we also believe that challenges breed opportunities. One of the strengths of weed science is that it brings together a broad base of scientific disciplines used in very pragmatic ways to produce solutions for end-users. If the weed science community perceives climate change and carbon dioxide as real challenges, then we can begin to meet the scientific and technical challenges needed to address the biotic threats posed by weeds. A formidable task, but within the scope of those of us that remain fascinated not only by plants, but by the plant pariahs of human society.

References

Ahmed FE, Hall AE, Madore MA. 1993. Interactive effects of high temperature and elevated carbon dioxide concentration on cowpea [(*Vigna unguiculata* (L.) Walp.]. *Plant Cell Environ.* **16**:835–842.

Ampong-Nyarko K, De Datta SK. 1991. *A Handbook for Weed Control in Rice*. IRRI Press, Los Banos, Philippines, 113 pp.

Archambault DJ. 2007. Efficacy of herbicides under elevated temperature and CO_2. In: Newton PCD, Carran A, Edwards GR, Niklaus PA (eds) *Agroecosystems in a Changing Climate*. CRC Press, Boston, MA, pp. 262–279.

Blumenthal D. 2005. Interrelated causes of plant invasion. *Science* **310**:243–244.

Blumenthal D. 2006. Interactions between resource availability and enemy release in plant invasion. *Ecol. Lett.* **9**:887–895.

Belote RT, Weltzin JF, Norby RJ. 2003. Response of an understory plant community to elevated CO_2 depends on differential responses of dominant invasive species and is mediated by soil water availability. *New Phytol.* **161**:827–835.

Bradley BA, Oppenheimer M, Wilcove DS. 2009. Climate change and plant invasion: restoration opportunities ahead? *Global Change Biol.* **15**:1511–1521.

Bridges DC. 1992. *Crop Losses Due to Weeds in the United States*. Weed Science Society of America, Champaign, IL, 403 pp.

Burden JJ, Elmquist T. 1996. Selective sieves in the epidemiology of *Melampsora lini*. Plant Path. **45**:933–943.

Canadell JG, Le Quéré C, Raupach MR, Field CB, Buttenhuis ET, Ciais P, Conway TJ, Gillett NP, Houghton RA, Marland G. 2007. Contributions to accelerating atmospheric CO_2 growth from economic activity, carbon intensity, and efficiency of natural sinks. *Proc. Natl. Acad. Sci. U. S. A.* **104**:18866–18870.

Cotton PA. 2003. Avian migration phenology and global climate change. *Proc. Natl. Acad. Sci. U. S. A.* **100**:12219–12222.

Crossman ND, Bryan BA, Bardsley DK. 2008. *Modelling Native and Exotic Flora Distributions Under Climate Change*. CSIRO Land and Water Science Report 01/08, ISSN: 1834–6618.

Department of Primary Industries (DPI). 2008. *Climate Change and Potential Distribution of Weeds*. Victorian Government Publication, Melbourne, Australia, 196 pp.

Dukes JS. 2000. Will the increasing atmospheric CO_2 Concentration affect the success of invasive species? In: Mooney HA, Hobbs RJ (eds) *Invasive Species in a Changing World*. Island Press, Washington, DC, pp. 95–113.

Dukes JS. 2002. Comparison of the effect of elevated CO_2 on an invasive species (*Centaurea solstitialis*) in monoculture and community settings. *Plant Ecol.* **160** (2): 225–234.

Dukes JS, Pontius J, Orwig D, Garnas JR, Rodgers VL, Brazee N, Cooke B, Theoharides KA, Stange EE, Harrington R, Ehrenfeld J, Gurevitch J, Lerdau M, Stinson K, Wick R, Ayres M. 2009. Responses of insect pests, pathogens, and invasive plant species to climate change in the forests of northeastern North America: what can we predict? *Can. J. For. Res.* **39**:231–248.

Fitter AH, Fitter RSR. 2002. Rapid changes in flowering time in British plants. *Science* **296**:1689–1691.

Fleming RA, Tatchell GM. 1995. Shifts in the flight responses of British aphids: a response to climate warming? In: Harrington R, Stork NE (eds.), *Insects in a Changing Environment*. Academic Press, London, pp. 505–508.

Gelbard JL, Belnap J. 2003. Roads as conduits for exotic plant invasions in a semiarid landscape. *Conserv. Biol.* **17**:420–432.

Gleadow RM, Evans JR, McCaffrey S, Cavagnaro TR. 2009. *Growth and Nutritive Value of Cassava (Manihot esculenta Cranz.) Are Reduced When Grown at Elevated* CO_2. Plant Biology Published Online, August 6, 2009. DOI: 10.1111/j.1438–8677.2009.00238.x.

Hattenschwiler S, Korner C. 2003. Does elevated CO_2 facilitate naturalization of the non-indigenous *Prunus laurocerasus* in Swiss temperate forests? *Funct. Ecol.* **17**:778–785.

IPCC: Climate Change. 2007. *Impacts, Adaptation and Vulnerability*. IPCC Secretariat, Geneva, Switzerland, 986 pp.

Karl TR, Melillo JM, Peterson TC. 2009. *Global Climate Change Impacts in the United States*. Cambridge University Press, MA, 188 pp.

Labrada R. 2007. The need for improved weed management in rice. In: *Proceedings of the 20th Session of the International Rice Commission*, Bangkok, Thailand, July 23–26, FAO, Rome, pp. 310–324.

Lawton JH. 2000. *Community Ecology in a Changing World*. InterResearch, Oldendorf/Luhe, Germany, 324 pp.

Mandersheid R, Weigel HJ. 1997. Photosynthetic and growth responses of old and modern spring wheat cultivars to atmospheric CO_2 enrichment. *Agric. Ecosys. Environ.* **64**:65–71.

Mohan JE, Ziska LH, Schlesinger WH, Thomas RB, Sicher RC, George K, Clark JS. 2006. Biomass and toxicity responses of poison ivy (*Toxicodendron radicans*) to elevated atmospheric CO_2. *Proc. Nat. Acad. Sci.* **103**:9086–9089.

Moya TB, Ziska LH, Namuco OS, Olszyk D. 1998. Growth dynamics and genotypic variation in tropical field grown paddy rice (*Oryza sativa* L.) in response to increasing carbon dioxide and temperature. *Global Change Biol.* **4**:645–656.

National Research Council. 2002. *Predicting Invasions of Non-indigenous Plants and Plant Pests*. National Academy Press, Washington, DC, 194 pp.

Newton PCD, Edwards GR. 2007. Plant breeding for a changing environment. In: Newton PCD, Carran RA, Edwards GR, Niklaus PA (eds) *Agroecosystems in a Changing Climate*. CRC Press, Boca Raton, FL, pp. 309–319.

Olfert O, Hallett R, Weiss RM, Soroka J, Goodfellow S. 2006. Potential distribution and relative abundance of swede midge *Contarinia nasturtii*, an invasive pest in Canada. *Entomol. Exp. Appl.* **120**:221–228.

Oliveira CM, Fontes JRA. 2008. Weeds as hosts for new crop pests: the case of *Protortonia navesi* (Hemiptera: Monophlebidae) on cassava in Brazil. *Weed Res.* **48**:197–200.

Parmesan C. 2007. Influences of species, latitudes and methodologies on estimates of phenological response to global warming. *Global Change Biol.* **13**:1860–1872.

Parmesan C, Ryrholm N, Stefanescu C, Hill JK, Thomas CD, Descimon H, Huntley B, Kaila L, Kullberg J, Tammaru T, Tennent J, Thomas JA, Warren M. 1999. Poleward shift of butterfly species' ranges associated with regional warming. *Nature* **399**:579–583.

Patterson DT. 1995. Effects of environmental stress on weed/crop interactions. *Weed Sci.* **43**:483–490.

Pimentel D, Zuniga R, Morrison D. 2005. Update on the environmental and economic costs associated with alien-invasive species in the United States. *Ecological Econ.* **52**:273–288.

Polley HW, Johnson HB, Mayeux HS. 1994. Increasing CO_2: comparative responses of the C_4 grass *Schizachyrium* and grassland invader *Prosopis*. *Ecology* **75**:976–988.

Raupach MR, Marland G, Ciais P, Quéré CL, Canadell JG, Klepper G, Field CB. 2007. Global and regional drivers of accelerating CO_2 emissions. *Proc. Natl. Acad. Sci. U. S. A.* **104**:18866–18870.

Robinson A, Robinson Z. 1997. Science has spoken, global warming is a myth. *The Wall Street Journal*, December 4, 1997.

Rosenberg JR, Burt PJA. 1999. Windborne displacements of desert locusts from Africa to the Caribbean and South America. *Aerobiologia* **15**:167–175.

Roy BA, Gusewell S, Harte J. 2004. Response of plant pathogens and herbivores to warming experiment. *Ecology* **85**:2570–2581.

Sasek TW, Strain BR. 1990. Implications of atmospheric CO_2 enrichment and climatic change for the geographical distribution of two introduced vines in the USA. *Clim. Change* **16**:31–51.

Scherm H, Coakley SM. 2003. Plant pathogens in a changing world. *Aust. J. Plant Pathol.* **32**:157–165.

Smith SD, Huxman TE, Zitzer SF, Charlet TN, Housman DC, Coleman JS, Fenstermaker LK, Seemann JR, Nowak RS. 2000. Elevated CO_2 increases productivity and invasive species success in an arid ecosystem. *Nature* **408**:79–82.

Song L, Wu J, Li C, Li F, Peng S, Chen B. 2009. Different responses of invasive and native species to elevated CO_2 concentration. *Acta Oecol.* **35**:128–133.

Stoll-Kleemann S, O'Riordan T, Jaeger CC. 2001. The psychology of denial concerning climate mitigation measures. *Global Environ. Change* **11**:107–117.

Sutherst RW, Baker RHA, Coakley SM, Harrington R, Kriticos DJ, Scherm H. 2007. Pests under global change—meeting your future landlords? In: Canadell JC, Pataki DE, Pitelka LF (eds) *Terrestrial Ecosystems in a Changing World.* Springer, Berlin, Heidelberg, pp. 211–226.

Time Magazine, 51 Ways to Save the Environment, April 9, 2007.

Treharne K. 1989. The implications of the "greenhouse effect" for fertilizers and agrochemicals. In: Bennet RC (ed.) *The Greenhouse Effect and UK Agriculture.* Ministry of Agriculture, Fisheries and Food, London, UK, pp. 67–78.

Tyser RW, Worley CA. 1992. Alien flora in grasslands adjacent to road and trial corridors in Glacier National Park, Montana (USA). *Conserv. Biol.* **6**:253–262.

United States Bureau of the Census (USBC). 1998. *Statistical Abstract of the United States*, 200th edn. U.S. Bureau of the Census, U.S. Government Printing Office, Washington, DC.

Walther G-R, Post E, Convey P, Menzel A, Parmesan C, Beebee TJC, Fromentin J-M, Hoegh-Guldberg O, Bairlein F. 2002. Ecological responses to recent climate change. *Nature* **416**:389–395.

Watson WA, Litovitz TL, Klein-Schwartz W, Rodgers GC, Youniss J, Reid N, Rouse WG, Rembert RS, Borys D. 2004. 2003 annual report of the American association of poison control centers toxic exposure surveillance system. *Am. J. Emerg. Med.* **22**:335–404.

Watt AD, Leather SR. 1986. The pine beauty in Scottish lodgepole pine plantations. In: Berryman AA (ed.) *Dynamics of Forest Insect Populations: Patterns, Causes, Implications.* Plenum Press, New York, pp. 243–266.

Wayne P, Foster S, Connolly J, Bazzaz FA, Epstein PR. 2002. Production of allergenic pollen by ragweed (*Ambrosia artemisiifolia* L.) is increased in CO_2-enriched atmospheres. *Ann Allergy Asthma Immunol.* **80**:669–679.

Westbrooks R. 1998. *Invasive Plants, Changing the Landscape of America: Fact Book.* Federal Interagency Committee for the Management of Noxious and Exotic Weeds (FICMNEW), Washington, DC, 109 pp.

Ye-Ebiyo Y, Pollack RJ, Spielman A. 2000. Enhanced development in nature of larval *Anopheles arabiensis* mosquitoes feeding on maize pollen. *Am. J. Trop. Med. Hyg.* **63**:90–93.

Zhang ZP. 2001. Weed management in rice in China. Summary presented at FAO Workshop on *Echinochloa* spp. Control, Beijing, China, May 27, 2001.

Ziska LH, Blumenthal DM. 2007. Empirical selection of cultivated oat in response to rising atmospheric carbon dioxide. *Crop Sci.* **47**:1547–1552.

Ziska LH, Bunce JA. 2000. Sensitivity of field-grown soybean to future atmospheric CO_2: selection for improved productivity in the 21st century. *Aust. J. Plant Physiol.* **27**:979–984.

Ziska LH, Emche SD, Johnson EL, George K, Reed DR, Sicher RC. 2005b. Alterations in the production and concentration of selected alkaloids as a function of rising atmospheric carbon dioxide and air temperature: implications for ethno-pharmacology. *Global Change Biol.* **11**:1798–1807.

Ziska LH, Faulkner SS, Lydon J. 2004. Changes in biomass and root:shoot ratio of field-grown Canada thistle (*Cirsium arvense*), a noxious, invasive weed, with elevated CO_2: implications for control with glyphosate. *Weed Sci.* **52**:584–588.

Ziska LH, Gebhard DE, Frenz DA, Faulkner S, Singer BD, Straka JG. 2003. Cities as harbingers of climate change: common ragweed, urbanization and public health, *J. Allergy Clin. Immunol.* **111**:290–295.

Ziska LH, George K. 2004. Rising carbon dioxide and invasive, noxious plants: potential threats and consequences. *World Res. Rev.* **16**:427–447.

Ziska LH, Reeves JB III, Blank R. 2005a. The impact of recent increases in atmospheric CO_2 on biomass production and vegetative retention of Cheatgrass (*Bromus tectorum*): implications for fire disturbance. *Global Change Biol.* **11**:1325–1332.

Ziska LH, Manalo PA, Ordonez RW. 1996. Intraspecific variation in the response of rice (*Oryza sativa* L.) to increased CO_2 and temperature: growth and yield response of 17 cultivars. *J. Exp Bot.* **47**:1353–1359.

Appendix

Common name	Scientific name	WSSA/ Bayer code	Brief description
Alfalfa*	*Medicago sativa* L.	MEDSA	Forage crop or volunteer weed
Alligatorweed	*Alternanthera philoxeroides* (Mart.) Griseb.	ALRPH	Invasive aquatic tropical weed
Anoda, spurred	*Anoda cristata* (L.) Schlecht.	ANVCR	Invasive annual agronomic weed
Arabidopsis	*Arabidopsis thaliana* (L.) Heynh.	ARBTH	Minor weed of mustard family
Asparagus*	*Asparagus officinalis* L.	ASPOF	Vegetable or volunteer weed
Barnyardgrass	*Echinochloa crus-galli* (L.) Beauv	ECHCG	Annual grass, weed in paddy rice
Barnyardgrass, smooth	*Echinochloa glabrescens* Munro ex Hook.	ECHGL	Annual grass, weed in paddy rice
Bermudagrass	*Cynodon dactylon* (L.) Pers.	CYNDA	Aggressive grass weed in lawns, fields
Bindweed, field	*Convolvulus arvensis* L.	CONAR	Rapidly growing climbing vine
Birdsfoot trefoil	*Lotus corniculatus* L.	LOTCO	Perennial legume, can be invasive
Blackberry, wild	*Rubus ursinus* L.	RUBUR	Perennial bramble bush
Broomsedge	*Andropogon virginicus* L.	ANOVI	Perennial weedy pasture grass
Brome, Red	*Bromus rubens* L.	BRORU	Invasive annual grass of western United States
Bulrush	*Scirpus* spp.	SCP??	Fast growing riparian sedge
Buttercup, creeping	*Ranunculus repens* L.	RANRE	Herbaceous, stoloniferous invasive
Buttercup, hairy	*Ranunculus sardous* Crantz	RANSA	Herbaceous forb common in eastern United States
Canada thistle	*Cirsium arvense* (L.) Scop.	CIRAR	Noxious invasive perennial of rangelands, farms
Cassia	*Cassia* spp.	CAS??	Tropical leguminous weed
Castor bean	*Ricinis communis* L.	RIICO	Poisonous hydrophilic weed

Common name	Scientific name	WSSA/ Bayer code	Brief description
Cheatgrass	*Bromus tectorum* L.	BROTE	Flammable invasive grass of rangelands
Chickweed, common	*Stellaria media* (L.) Vill.	STEME	Annual weed of lawns, landscapes
Cinquefoil, sulfur	*Potentilla recta* L.	PTLRC	Unpalatable weed of rangelands
Clover, red	*Trifolium pretense* L.	TRFPR	Perennial weed of pasture and forage crops
Clover, white	*Trifolium repens* L.	TRFRE	Shallow rooted winter perennial legume
Coca	*Erythoxylem coca* Lam.	ERYCO	Perennial herb, narcotic
Cocklebur	*Xanthium strumarium* L.	XANST	Herbaceous annual riparian weeds
Cogongrass	*Imperata cylindrica* (L.) Beauv.	IMPCY	Tropical invasive rhizomatous grass
Coontail, common	*Ceratophyllum demersum* (Cham.) Asch.	CERDE	Common water weed
cordgrass, prairie	*Spartina pectinata*	SPTPE	Native riparian grass
Corn*	*Zea mays* L.	ZEAMX	Tropical grass grown for food, feed, and fuel
Corn cockle	*Agrostemma githago* L.	AGOGI	Annual forb, cereal weed
Cotton	*Gossypium hirsutum* L.	GOSHI	Crop for fiber
Crabgrass, large	*Digitaria sanguinalis* (L.) Scop.	DIGSA	Annual lawn and forage weed
Crabgrass, smooth	*Digitaria ischaemum* (Schreb) ex Muhl.	DIGIS	Annual lawn and forage weed
Crabgrass, southern	*Digitaria ciliaris* (Retz.) Koel.	DIGSP	Annual lawn and forage weed in Southeast
Dandelion	*Taraxacum officinale* Weber	TAROF	Ubiquitous lawn weed
Diviners Sage, Salvia	*Salvia divinorum Epling* & *Játiva*	SADIV	Psychoactive herb
Dock, curly	*Rumex crispis* L.	RUMCR	Riparian broadleaf perennial weed
Dock, broadleaf	*Rumex obtusifolius* L.	RUMOB	Taprooted perennial similar to curly dock
Duckweed, common	*Lemna minor* L.	LEMMI	Small aquatic weed
English ivy	*Hedera helix* L.	HEEHE	Invasive vine of woodlands
European beachgrass	*Ammophila arenaria* L. (link)	AMMAR	Invasive grass of western United States and Canada
Foxtail, giant	*Setaria faberi* Herrm.	SETFA	Exotic annual grass of mid-west United States
Foxtail, yellow	*Setaria glauca* (L.) Beauv.	SETLU	Ubiquitous annual grass of continental United States
Foxtail, green	*Setaria viridis* (L.) Beauv.	SETVI	Ubiquitous annual grass of continental United States
Garlic mustard	*Alliaria petiolata* Andrz.	ALAPE	Invasive biennial forb of woody areas
Garlic, wild	*Allium vineale* L.	ALLVI	Invasive perennial weed of mid-Atlantic

Common name	Scientific name	WSSA/ Bayer code	Brief description
Giant Hogweed	*Heracleum mantegazzianum* Somm. & Lev.	HERMA	Invasive biennial with phytotoxic sap
Giant Reed, Arundo	*Arundo donax* L.	ARUDO	Invasive riparian weed of western United States
Golden bristlegrass	*Setaria sphacelata* var. anceps	SETSP	Pasture grass, can be volunteer weed
Goosegrass	*Eleusine indica* (L.) Gaertn.	ELEIN	Annual agronomic weed
Groundsel, common	*Senecio vulgaris* L.	SENVU	Winter/Spring annual weed of landscapes
Guineagrass	*Panicum maximum* Jacq.	PANMA	Perennial invasive grass of tropics/subtropics
Gumweed, curlycup	*Grindelia squarrosa* L.	GRNSQ	Xeric weed of western United States
Hawkweed, meadow	*Hieracium caespitosum* Dumort.	HIECA	Invasive perennial of northwest United States
Hawkweed, orange	*Hieracium aurantiacum* L.	HIEAU	Invasive perennial, especially of upper Midwest United States
Hoary cress	*Cardaria draba* (L.) Desv	CADDR	Invasive perennial forb of rangelands
Honey mesquite	*Prosopis glandulosa* Torr.	PRCJG	Noxious perennial xeric weed
Horsetail	*Equisetum arvense* L.	EQUAR	Rhizomatous perennial of landscapes
Horseweed	*Conyza canadensis* (L.) Cronq.	ERICA	Annual weed of disturbed areas
Houndstongue	*Cynoglossum officinale* L.	CYWOF	Poisonous biennial noxious weed
Itchgrass	*Rottboellia cochinchinensis* (Lour.) W. Clayton	ROOEX	Annual, invades disturbed and agronomic areas
Japanese honeysuckle	*Lonicera japonica* Thunb.	LONJA	Invasive forest vine of eastern United States
Japanese millet*	*Echinochloa esculenta* (A. Braun) H. Scholz	ECHES	Forage crop, volunteer weed
Japanese stiltgrass	*Microstegium vimineum* Trin Camus	MCVGM	Annual invasive grass, especially woody areas
Johnson grass	*Sorghum halepense* (L.) Pers.	SORHA	Exotic perennial, aggressive, wide distribution
Jointed goatgrass	*Aegilops cylindrical* Host	AEGCY	Winter annual found in cereals and rangelands
Knapweed, spotted	*Centaurea maculosa* Lam.	CENMA	Invasive weed of western rangelands
Knapweed, diffuse	*Centaurea diffusa* Lam.	CENDI	Invasive weed of western rangelands
Knotweed, common	*Polygonum plebeium* R. Br.	POLPL	Weed of trampled and compacted areas
Knotweed, Japanese	*Fallopia japonica* (Houtt.) Ronse Decr.	POLCU	Invasive perennial of riparian systems

Common name	Scientific name	WSSA/ Bayer code	Brief description
Kudzu	*Pueraria montana* var. *lobata* (Willd.)	PUELO	Highly invasive vine of southeastern United States
Ladysthumb	*Polygonum persicaria* L.	POLPE	Much branched summer annual weed
Lambsquarters, common	*Chenopodium album*	CHEAL	Ubiquitous annual, common agronomic weed
Larkspur	*Delphinium* spp.	DEL??	Numerous spp. Poisonous rangeland weed
Leafy spurge .	*Euphorbia esula* L.	EPHHL	Invasive perennial, especially in no-till agriculture
Little bluestem	*Schizachyrium scoparium* (Michx.) Nash	ANOSC	Native prairie grass
Loosestrife, purple	*Lythrum salicaria* L.	LYTSA	Invasive perennial of riparian areas
Loosestrife, gooseneck	*Lysimachia clethroides*	LYCLE	Perennial, less widespread than *L. salicaria*
Marestail	*Hippuris vulgaris* L.	HPPVU	Perennial aquatic herb, native to California
Melaleuca (punk) tree	*Melaleuca quinquenervia* (Cav.) S.F. Blake	MEQUI	Invasive tree of Florida, Louisiana
Milkweed, common	*Asclepias syriaca* L.	ASCSY	Perennial agronomic weed, especially in no-till
Millet, wild-proso*	*Panicum miliaceum* L.	PANMI	Weed or forage grass
Morningglory, pitted	*Ipomoea lacunose* L.	IPOLA	Native perennial vine of eastern United States
Morningglory, bigroot	*Ipomoea pandurata* (L.) G.F.W. Meyer	IPOPA	As *L. lacunose*, but more in southeastern United States
Morningglory, ivyleaf	*Ipomoea hederacea* (L.) Jacq.	IPOHE	Annual vine of disturbed areas
Morningglory, three lobe	*Ipomoea triloba* L.	IPQTR	Perennial subtropical vine, especially Florida
Morningglory, tall	*Ipomoea purpurea* (L.) Roth	PHBPU	Ubiquitous annual vine, especially cultivated fields
Mustard, black	*Brassica nigra* (K.) W.J.D. Koch	BRSNA	Annual weed widespread within disturbed areas
Mustard, birdsrape	*Brassica rapa* L.	BRSRA	Annual/biennial forb throughout North America
Mustard, wild	*Brassica kaber* (DC.) L.C. Wheeler	SINAR	Annual weed in forage and small grains
Nettle, stinging	*Urtica dioica* L.	URTDI	Exotic perennial, induces contact dermatitis
Nettle, burning	*Urtica urens* L.	URTUR	Annual, induces contact dermatitis
Nightshade	*Solanum* spp.	SOL??	Number of spp., highly poisonous
Nightshade, American black	*Solanum nigrum* L. var. americanum	SOLAM	Annual/perennial poisonous weed, especially South

Common name	Scientific name	WSSA/ Bayer code	Brief description
Nightshade, bittersweet	*Solanum elaeagnifolium* Cav.	SOLEL	Perennial vine, often on moist soils
Nightshade, black	*Solanum nigrum* L.	SOLNI	Annual, sometimes perennial agronomic weed
Nightshade, West Indian	*Solanum ptycanthum* Dun.	SOLPT	Widely distributed annual forb
Nightshade, hairy	*Solanum sarrachoides* Sendtner	SOLSA	Common weed in potato, host for late blight
Nutsedge, purple	*Cyperus rotundus* L.	CYPRO	Perennial ubiquitous weed, especially agricultural fields
Nutsedge, yellow	*Cyperus esculentus* L.	CYPES	Perennial widely distributed weed
Oat, wild	*Avena fatua* L.	AVEFA	Annual weed, especially cereals
Onion, wild, oniongrass	*Allium canadense* L.	ALLCO	Perennial forb, widely distributed in landscapes
Orchardgrass*	*Dactylis glomerata* L.	DACGL	Perennial forage grass, or volunteer weed
Oriental bittersweet	*Celastrus orbiculatus* Thunb.	CELOR	Invasive woody vine of northeastern United States
Panicum, broadleaf	*Panicum adspersum* Trin.	PANAD	Annual grass, found in Florida
Panicum, fall	*Panicum dichotomiflorum* Michx.	PANDI	Widely distributed annual weedy grass
Panicum, browntop	*Panicum fasciculatum* Sw.	PANFA	Annual weed of southern United States
Panicum, Texas	*Panicum texanum* Buckl.	PANTE	Annual, southeastern United States especially agricultural areas
Peyote	*Lophophora williamsii* (Lem.) J. Coult.	LOPWI	Psychotropic cactus
Pigweed	*Amaranth* spp.	AMA??	Troublesome weedy spp. widely distributed
Pigweed, Palmer	*Amaranthus palmeri* S. Wats.	AMAPA	Aggressive annual weed, especially southern United States
Pigweed, Powell	*Amaranthus powellii* S. Wats.	AMAPO	Annual weed, widely distributed
Pigweed, prostrate	*Amaranthus blitoides* S. Wats.	AMABL	Annual weed, widely distributed
Pigweed, redroot	*Amaranthus retroflexus* L.	AMARE	Annual agronomic weed, widely distributed
Pigweed, spiny	*Amaranthus spinosus* L.	AMASP	Noxious annual weed, especially eastern United States
Pigweed, tumble	*Amaranthus albus* L.	AMAAL	Widely distributed annual weed
Plantain, narrowleaf	*Plantago lanceolata* L.	PLALA	Common weed, especially of lawns/landscapes
Plantain, broadleaf	*Plantago major* L.	PLAMA	Common weed, especially of lawns/landscapes

Common name	Scientific name	WSSA/ Bayer code	Brief description
Poison hemlock	*Conium maculatum* L.	COIMA	Poisonous weed of disturbed areas
Poison ivy	*Toxicodendron radicans* (L.) Ktze.	TOXRA	Widely distributed vine. Induces dermatitis
Pokeweed	*Phytolacca americana* L.	PHTAM	Perennial summer weed, especially disturbed sites
Poppy, opium	*Papaver somniferum* L.	PAPSO	Annual, source of opiates morphine, heroin
Poppy, wild	*Papaver setigerum* DC	PAPSE	Annual, weed in commercial poppy fields
Potato*	*Solanum tuberosum* L.	SOLTU	Tuber crop, occasional volunteer weed
Prickly pear cactus	*Opuntia* spp.	OPU??	Invasive cactus of Australia
Prickly sida	*Sida spinosa* L.	SIDSP	Ubiquitous summer annual, especially in cultivation
Princess tree	*Paulownia tomentosa* (Thunb.)	PAZTO	Invasive tree from China
Puncturevine	*Tribulus terrestris* L.	TRBTE	Prostrate weed, spiny seeds, widely distributed
Purslane, common	*Portulaca oleracea* L.	POROL	Prostrate summer annual, widely distributed
Quackgrass	*Elytrigia repens* (L.) Nevski	AGRRE	Invasive perennial grass, widely distributed
Radish, wild	*Raphanus raphanistrum* L.	RAPRA	Annual weed, especially of forage crops
Ragweed, common	*Ambrosia artemisiifolia* L.	AMBEL	Annual weed, especially disturbed areas
Ragweed, giant	*Ambrosia trifida* L.	AMBTR	Annual, prefers moist soil, agronomic weed
Ragweed, western	*Ambrosia psilostachya* L.	AMBPS	Erect perennial, especially in waste areas
Rhubarb	*Rheum Rhabarbarum* L.	RHURH	Leafy vegetable
Rice*	*Oryza sativa* L.	ORYSA	Global cereal or, if wild, volunteer weed
Rocket, yellow	*Barbarea vulgaris* R. Br.	BARVU	Winter annual /biennial, especially in turfgrass
Rosary pea	*Abrus precatorius* L.	ABRPR	Poisonous invasive of Florida
Rush skeleton weed	*Chondrilla juncea* L.	CHOJU	Invasive perennial of rangelands
Russian thistle	*Salsola iberica* Sennen & Pau	SASKR	Numerous spp. widely distributed annual
Ryegrass, Italian	*Lolium multiflorum* L.	LOLMU	Annual weed, especially cool-season crops
Ryegrass, perennial	*Lolium perenne* L.	LOLPE	Perennial weed, widely distributed

Common name	Scientific name	WSSA/ Bayer code	Brief description
Sagebrush, common	*Artemesia tridentata*, var. subsp.	ARTTR	Woody perennial of western United States
Salt cedar	*Tamarix ramosissima*	TAARA	Invader of riparian systems
Sesbania, hemp	*Sesbania exaltata* (Raf.)	SEBEX	Riparian weed of disturbed sites
Shepherd's purse	*Capsella bursa-pastoris* (L.) Medicus	CAPBP	Ubiquitous winter annual weed
Sandbur	*Cenchrus longispinus*	CCHPA	Lawn weed, sharp spines
Showy crotalaria	*Crotalaria spectabilis* Roth	CVTSP	Annual agronomic weed, especially in southern United States
Sicklepod	*Cassia obtusifolia* L.	CASOB	Agronomic weed, especially soybean, peanut
Sorghum/ shattercane*	*Sorghum bicolor* (L.) Moench.	SORVU	Cereal grain or volunteer weed
Sowthistle, annual	*Sonchus oleraceus* L.	SONOL	Widely distributed agronomic annual weed
Sowthistle, perennial	*Sonchus arvensis* L.	SONAR	Perennial weed of disturbed areas
Soybean	*Glycine max* L. Merr.	GLYMA	Commercial annual crop
St. Johns wort	*Hypericum perforatum* L.	HYPPE	Noxious weed, toxic to livestock
Sugarbeet	*Beta vulgaris* L.	BETVU	Annual crop, commercial sugar source
Sunflower*	*Helianthus annuus* L.	HELAN	Oil-seed crop, volunteer weed
Tansy ragwort	*Senecio jacobaea* L.	SENJA	Invasive weed of Western rangelands
Thistle, musk	*Carduus nutans* L.	CRUNU	Biennial xeric weed of disturbed areas
Thistle, bull	*Carduus vulgare* (Savi) Tenore	CIRVU	Spiny biennial weed, widely distributed
Toadflax, Dalmatian	*Linaria dalmatica* (L.) Mill.	LINDA	Invasive perennial weed of rangelands
Tree of heaven	*Ailanthus altissima* (Mill.) Swingle	AILAL	Urban invasive perennial
Velvetleaf	*Abutilon theophrasti* Medicus	ABUTH	Opportunistic agronomic annual weed
Vetch, common	*Vicia sativa* L.	VICSA	Annual weed, can be used as "green manure"
Vetch, hairy	*Vicia villosa* Roth	VICVI	Annual weed, can be cover crop
Water hemlock	*Cicuta maculate*	CIUMC	Poisonous perennial weed
Water hyacinth	*Eichhornia crassipes* (Mart.) Solms	EICCR	Aggressive aquatic weed
Water thyme, hydrilla	*Hydrilla verticillata* (L. f.) Royle	HYLLI	Invasive aquatic weed, especially southern United States
Wheat*	*Triticum aestivum* L.	TRZAW	Cereal grain, occasional volunteer weed
White snakeroot	*Ageratina altissima* (L.) King & H. Rob.	AGALT	Poisonous perennial herb

Common name	Scientific name	WSSA/ Bayer code	Brief description
Witchgrass	*Panicum capillare* L.	PANCA	Ubiquitous annual weed
Wooly cupgrass	*Eriochloa villosa* (Thunb.) Kunth	ERBVI	Annual agronomic weed of midwestern United States
Yellow salsify	*Tragopogon dubius* Scop.	TRODM	Widespread invasive annual weed
Yellow star thistle	*Centaurea solstitialis* L.	CENSO	Annual invasive weed, especially western United States

* Can be crop or weed.

Index

Printed and bound by CPI Group (UK) Ltd, Croydon, CR0 4YY
16/09/2026
14963449-0001